Gold, Platinum, Palladium, Silver & Other Jewelry Metals

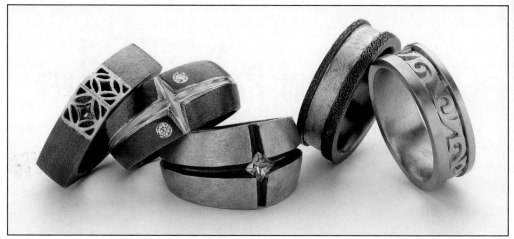

Rings made of 18K yellow or rose gold, palladium and oxidized sterling silver. *Jewelry by Alishan; photo by John Parrish.*

Stainless Damascus steel rings with ceramic inlays and black oxide. *Jewelry by Chris Ploof Designs; photo by Robert Diamante.*

Title page: Bracelet designed by Peter Schmid made of oxidized sterling silver and 24K and 22K gold set with a 5.08-carat amethyst. *Jewelry and photo courtesy Atelier Zobel in Germany; photo by Fred Thomas.*

Left: Wedding bands by Chris Nelson of Urban-Armour Jewelry, made of 22K gold fused onto hand-textured iron and set with "raw" uncut diamonds. *Photo: Chris Nelson / Urban-Armour.*

GOLD, PLATINUM
PALLADIUM, SILVER
& OTHER JEWELRY METALS

How to Test, Select & Care For Them

Renée Newman

International Jewelry Publications

Los Angeles _____

International Jewelry Publications
P.O. Box 13384
Los Angeles, CA 90013-0384 USA

(Inquiries should be accompanied by a self-addressed, stamped envelope.)

Printed in Singapore

Library of Congress Cataloging-in-Publication Data

Newman, Renée.
  Gold, platinum, palladium, silver & other jewelry metals: how to test, select & care for them / Renee Newman.
     p. cm.
  Includes bibliographical references and index.
  ISBN 978-0-929975-47-4 (alk. paper)
1. Precious metals. 2. Art metal-work. 3. Jewelry making. 1. Title.
  TS729.N4825 2013
  739.27–dc23

                                                2012036077

Cover photos:

Mokume gane ring by James Binnion made of silver, palladium white gold and 18K gold. *Photo by Hap Sakwa.*

Chrysanthemum 18K gold and zircon ring by Alexandra Hart. *Photo by Ralph Gabriner.*

Palladium and 18K gold spessartine ring and photo by Erik Stewart.

Iron & 18K gold ruby pendant & photo by Chris Nelson of Urban-Armour.

Fire colored titanium dial inlaid with 24K gold & hand-engraved with portraits of the jeweler's daughters. *Dial, case & photo by James Roettger.*

Palladium and oxidized silver cuff by Todd Reed. *Photo by Brian Mark.*

Back cover photo: Mokume gane rings by James Binnion made of silver, palladium white gold and 18K gold. *Photo by Hap Sakwa.*

Spine photo: 24K earring by Yossi Harari. *Photo from Muse Imports.*

Contents

Acknowledgments

I would like to express my appreciation to the following people and companies for their contribution to *Gold, Platinum, Palladium, Silver & Other Jewelry Metals*:

Ernie and Regina Goldberger of the Josam Diamond Trading Corporation. This book could never have been written without the experience and knowledge I gained from working with them.

Eve Alfillé, Michael Allchin, Sandy Boothe, Dellana, Elaine Ferrari Santhon, Kim Fox, Miranda Hayes Schulz, Alan Hodgkinson, Thomas Jansseune, Doug Kerns, Jurgen Maerz, Dippal Manchanda, Lindy L Matula, Mark B. Mann, Patara Marlow, Steve Mikaelian, Chris Nelson, Danusia V. Niklewicz, Kate Peterson, James Poag, Pat Pruitt, Michael Radomyshelsky, Alan Revere, Bob Romanoff, Marc Rothenberg, Debra Sawatsky, Kathrin Schoenke, Bill Seeley, Spider, Hanna Cook-Wallace, William B. Whetstone. They've made valuable suggestions, corrections and comments regarding the portions of the book they examined. They're not responsible for any possible errors, nor do they necessarily endorse the material contained in this book.

A & A Findings, Gelin Abaci, Mark Fischer, Josam Diamond Trading Corp, Jane Keller, Jane Nordvedt, Pure Gold, Timeless Gem Designs, Varna, Sandra Weaver and Nerses Yahiayan. Jewelry and/or related materials from them were loaned for some of the photographs.

Adin Fine Antique Jewelry, Alan Hodgkinson, Alexandra Hart, Alishan, Arkenstone, Atelier Zobel, Belle Etoile Jewelry, Bobby Mann, Brian Eburah, Carina Rossner, Charles Lewton-Brain, Chris Nelson, Chris Ploof Designs, Craig O'Donnell, Crown-Ring, David Anderson, Debra Sawatzky, Dellana, Denney Jewelers, Erik Stewart, Eve J. Alfillé, Evenheat Kiln Company, Exotica Jewelry, FDJ On Time LLC, Frank Heiser, Fred & Kate Pearce, Frédéric DuClos, Fred Rich Enamel Design, Gemvision, GoldMoney, Gurhan, Handfast Design, Hanna Cook-Wallace, Helene M. Apitzsch, Holland Family Trust, Holly Gage, Hubert Inc, Intercept Silver and Jewelry Care Company, Jackie Truty, James Binnion, James Roettger, John S. Brana, Katy Briscoe, Kim Fox, Krikawa, Kristen Anderson, Lang Antique & Estate Jewelry, Lashbrook Designs, Lisa Stockhammer, Louise Walker, Mark Mann, Mark Schneider, Metal Marketplace International, Michael Zobel, Mr. Titanium, Olympus NDT, Pala International, Palladium Alliance International (PAI), Pat Pruitt, Platinum Guild International (Germany), Poag Jewellers, R. H. Wilkins Engravers Ltd, Rio Grande, Rob Lavinsky, Romanoff International Supply Corp, Rubin & Son, Sara Commers, Skinner Inc, Spider, Solidscape, So Young Park Stamper Black Hills Gold Jewelry,

Stephen Vincent, Stuller, Inc., The Roxx Limited, Three Graces Antique Jewelry, Todd Reed, TRI Electronics, True Knots, Urban-Armour Jewelry, Varna, Wayne Victor Meeten, William Newton, Xcalibur XRF Service, Yossi Harari. Photos or diagrams from them have been reproduced in this book.

Frank Chen and Joyce Ng. They have provided technical assistance.

Louise Harris Berlin. She has spent hours carefully editing *Gold, Platinum, Palladium, Silver & Other Jewelry Metals*. Thanks to her, this book is much easier for consumers to read and understand.

My sincere thanks to all of these contributors for their kindness and help.

1

Key Facts

Between 1990 and 2005, gold prices generally ranged between $275–$425 an ounce. Gold was the most important metal for fine jewelry, although platinum was becoming a popular choice for bridal jewelry. By the end of 2005, gold prices had risen above $500/oz and later skyrocketed. They reached $1895/oz in September 2011, and while prices a year later had declined somewhat, they still remained higher than those of platinum and therefore unaffordable for many people. As a result, silver and palladium have become increasingly popular for fine jewelry. Even nonprecious metals and alloys such as stainless steel, titanium, cobalt chrome and iron are being used for wedding jewelry. Consumers, jewelers and designers now have more options. This book will help you identify the various metals and understand their advantages and disadvantages.

When selecting jewelry, you should consider the type of metal and alloy used because it affects the value, durability and requirements for care. This chapter explains basic terminology used to describe metals, weights, measures and marks. Four price charts are included below to show you how the prices of gold, platinum, palladium and silver have fluctuated the last twenty years. Values and preferences change with time.

Fig. 1.1 Iron and 18K gold wedding bands & photo by Chris Nelson of Urban-Armour.

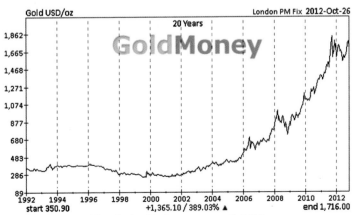

Fig 1.2 A 20-year historical gold chart courtesy GoldMoney.

Fig 1.3 A 20-year historical platinum chart courtesy GoldMoney.

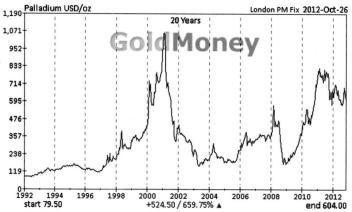

Fig 1.4 A 20-year historical palladium chart courtesy GoldMoney.

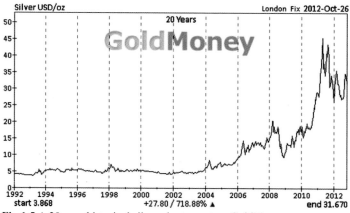

Fig 1.5 A 20-year historical silver chart courtesy GoldMoney.

Terms Related to Metals

Alloy: A mixture of two or more metals made by melting them together. Gold, for example, is alloyed (combined) with metals such as silver, copper, zinc and nickel to reduce its cost and change characteristics such as its color and hardness. **Platinum alloys** are usually made by combining platinum with ruthenium, palladium, cobalt, iridium or copper.

Base metal: A nonprecious metal such as copper, nickel or zinc

Pewter: A tin alloy. According to the US Federal Trade commission, pewter consists of at least 90 percent tin. In colonial America, pewter was an alloy of tin and lead. Today in England, "pewter" generally refers to a tin alloy made of 91% tin, 7% antimony and 2% copper. Elsewhere it may contain other elements such as lead, bismuth and zinc, and the percentages may vary.

Platinum group metals (PGM): Platinum and the five other metals which are chemically and physically similar and which are often deposited with it: palladium, iridium, osmium, rhodium and ruthenium. Platinum alloys are usually made by combining platinum with ruthenium, palladium, cobalt, iridium or copper.

Precious metals: Gold, silver and the platinum group metals. They are investment commodities with an inherent currency value as well as jewelry and industrial metals.

Tensile strength: A metal's ability to withstand the stress of stretching. It is determined by the amount of force required to break a specified unit area of a wire of the material. Platinum alloys are generally stronger than white gold, which in turn is normally stronger than sterling silver. Strength is different than hardness, which is a material's ability to withstand scratching. Gold, silver and platinum are alloyed to make them harder and more resistant to wear.

Table 1.1 lists the Mohs hardness of some pure metals to help you compare it to the hardness of gemstones. The Vickers and Brinell hardness systems are more commonly used for metals because of their increased degree of accuracy.

Table 1.1 Mohs hardness of *pure* metals

Metal	Mohs hardness	Metal	Mohs hardness
Silver	2.5 to 3	Niobium	6
Gold	2.5 to 3	Rhodium	6
Copper	3	Titanium	6
Iron	4	Iridium	6.5
Platinum	4 to 4.5	Ruthenium	6.5
Palladium	4.75	Tungsten	7.5

The above Mohs hardness values are from Wikipedia and the American Federation of Mineralogical Societies. Hardness tables of pure metals can be misleading. Most jewelry is made of alloys, which are normally harder than pure metals. Alloys vary in hardness and strength depending on the type and proportion of alloying elements and whether the metals have been work hardened or not. The Vickers and Brinell hardness scales are more commonly used for metals, but the table above allows you to compare metals to gems.

Terms Related to Gold Content

Fine gold: Gold containing no other elements or metals. It's also called pure gold or 24K (24 karat) gold and has a fineness of 999.

Fineness: The amount of gold or platinum in relation to 1000 parts. For example, gold with a fineness of 750 has 750 parts (75%) gold and 250 parts of other metals. An alloy containing 95% platinum and 5% ruthenium has a fineness of 950.

Karat (Carat): A measure of gold purity. One karat is 1/24 pure, so 24 karat is pure gold. Do not confuse "karat," the unit of gold purity, with "carat," the unit of weight for gemstones. These two words originate from the same source, the Italian *carato* and the Greek *karation* which mean "fruit of the carob tree." In ancient times, carob beans were used as counterweights when weighing gems and gold. Outside the US, "karat" is often spelled "carat," particularly in countries which are members of the British Commonwealth.

Karat gold: A gold alloy, which in the United States must have a fineness of at least 10K. In the UK, Spain, Japan and Canada, it must be at least 9K and in Mexico and Germany 8K. Gold must be at least 18K in France and Italy to be called gold. For detailed information on US government gold and jewelry terminology regulations see www.ftc.gov/bcp/guides/jewel-gd.htm. Chapter 2 has additional information.

Plumb gold (KP): In general usage, it means gold that has the same purity as the mark stamped on it. Therefore, 14KP means gold that is really 14K. Prior to 1978 in the US, 13K gold jewelry which had been soldered could be stamped 14K. When jewelers describe their jewelry as plumb gold, they are emphasizing it has at least the karat value that is marked and that they abide by the law because not all jewelers do.

Pure gold: Same as fine gold

Solid gold: Gold that is not hollow. Even though legally in the US, "solid gold" can only be used for 24K gold, it also refers to karat gold which is not hollow or layered.

Vermeil: Sterling silver covered with at least 120/millionths of an inch of fine gold. The layer of gold may be either electroplated or mechanically bonded. See Chapter 12.

Weights, Measures and Marks

Avoirdupois weight: The weight system used in the U.S. for food and people and almost everything except precious metals and gems. One avoirdupois pound equals 16 ounces.

Carat: A unit of weight for gems, which was standardized internationally in 1913 and adapted to the metric system, with one carat equaling 1/5 of a gram. The term "carat" sounds more impressive and is easier to use than fractions of grams. Consequently, it is the preferred unit of weight for gemstones.

Grain: A measurement of weight equaling 1/24 of a pennyweight. This was one of the earliest units of weight for gold. It was originally the equivalent of one grain of wheat.

Gram: The most widespread unit of weight for gold jewelry. See Table 1.2 for equivalent weights.

Pennyweight: Unit of weight equaling 1/20 of a troy ounce. In the Middle Ages it was the weight of a silver penny in Britain. Now pennyweight is used mainly in the American jewelry trade.

Hallmark: A group of official marks stamped on gold, silver, platinum or palladium objects to indicate their quality, maker and origin. The term refers to the Goldsmith's Hall in London, which has overseen the marking of gold in England since 1300. Hallmarking systems are found in European countries such as Belgium, France, Britain, Germany, Holland, Italy and Denmark. Figure 1.6 is an example of a British hallmark.

Fig. 1.6 Full set of English hallmarks with makers mark of "A & W" (for Alfred Wilcox Jewellers, first registered in 1880 and closed in 1959), 18K gold marks, Birmingham Assay Office mark (anchor), and date letter "h" for 1907. *Jewelry and photo courtesy Lisa Stockhammer-Mial of The Three Graces Antique Jewelry.*

Fig. 1.7 The suffragette ring on which the above hallmark was stamped. The term "suffragette" was coined by the British press about 1906 for jewelry pieces that were worn to advocate women's rights, including the right to vote. The color scheme, i.e., purple for dignity, white for purity and green for hope has been regarded as synonymous with the women's rights movements in both Europe and the United States. The triple play of white, green and purple also influenced fashion during the time, and not all pieces with this triad indicated the suffragette movement. *Ring and photo: The Three Graces Antique Jewelry.*

"Hallmark" also has a more general meaning which refers to any mark stamped on an article of trade to indicate origin, purity or genuineness. For example, the Japanese flag stamped on the 24K gold clasp in figure 1.9 signifies that the piece was inspected and meets Japanese standards for the quality mark of 1000 (24K gold).

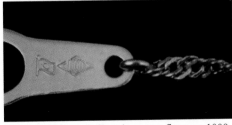

Fig. 1.8 24K—24 karat gold mark. *Photo © by Renée Newman.*

Fig. 1.9 Japanese hallmark next to fineness 1000 (quality) mark for the same chain as in figure 1.8. *Photo © by Renée Newman.*

Quality mark: A set of numbers, letters, or symbols stamped on metal to indicate its type and content (figs. 2.1 to 2.13). For example 18K means 75% gold, 900 Plat. means 90% platinum. In the US, jewelry which does not cross state lines has not been required to have a quality mark. Fineness and karat marks are quality marks. Additional photo examples are shown in the Chapters 2-5 on gold, platinum, palladium and silver.

Tael: Chinese gold weight. 1 tael=1.2034 ounces troy of fine gold.

Trademark: A mark that indicates the manufacturer, importer or seller of an item (whoever stands behind its quality mark) (figs. 2.3 & 2.4). In the USA, trademarks must be registered with the Patent and Trademark Office, and trademarked items must have a quality mark. In addition, any item that bears a quality mark should have a U.S. registered trademark. There is little enforcement of this law, however. Consequently, many types of jewelry marked 10K, 14K or 18K are not trademarked.

Fig. 1.10 "PLAT" for platinum 950, "750" for 18K gold and Varna Trademark. *Photo © by Renée Newman.*

Fig. 1.11 Platinum 900 fineness mark and trademark of Gelin & Abaci. *Photo © by Renée Newman..*

Fig 1.12 Fineness mark for 14K gold—585

Fig. 1.13 Fineness mark For 10K—417.

Fig 1.14 "18KP" means 18K plumb gold, not plated gold.

Troy ounce: The standard unit of weight for gold. It may have been named after a weight used at the annual fair at Troyes in France during the Middle Ages.

Specific gravity (SG): A measure that indicates the relative density and weight of a substance. It's the ratio that compares the weight of a substance to the weight of an equal volume of water at 4°C (39°F). Pure platinum, for example, has an SG of 21.4, meaning that it's 21.4 times heavier than an equal volume of water. Pure gold has an SG of 19.3, but when it's alloyed with 25% palladium, its SG decreases to 15.6. Pure silver has an SG of 10.6.

Troy weight: The system of weights used in the United States and United Kingdom for gold and silver in which one pound equals twelve ounces and one ounce equals twenty pennyweights. It should not be confused with avoirdupois weight.

Table 1.2 Weight Conversion Table

Unit of weight	Converted weight
1 pennyweight (dwt)	= 1.555 g = 0.05 oz t = 0.055 oz av = 7.776 cts
1 troy ounce (oz t)	= 31.103 g = 1.097 oz av = 20 dwt = 155.51 cts
1 ounce avoirdupois (oz av)	= 28.3495 g = 0.911 oz t = 18.229 dwt = 141.75 cts
1 carat (ct)	= 0.2 g = 0.006 oz t = 0.007 oz av = 0.1286 dwt
1 gram (g)	= 5 cts = 0.032 oz t = 0.035 oz av = 0.643 dwt

Miscellaneous Terms

Findings: Metal components used for jewelry construction or repair such as clasps, settings, studs and safety chains.

Hardening: The process of raising a metal's durability and resistance to denting or abrasion by means of cold working or by heat treating.

Mounting: The metal part of a jewelry piece before the stones are set into it.

Oxidation: A term meaning to combine with oxygen.

Refining: The process of separating pure precious metals from impurities and base metals.

Scrap: The metal by-product of fabrication that can be reclaimed and recycled by refining and/or smelting. A **scrapper** is a facility that refines and smelts scrap.

Shank: The part of a ring that encircles the finger and is attached to the stone setting(s).

Solder: A metal or metallic alloy used to join metals. Solder is designed to melt at a lower temperature than the metal to be joined. The terms **easy, medium**, and **hard solders** are used to describe solders with progressively higher melting points. Normally, hard solder is first used on a piece since it melts at the highest temperature. Medium and then easy solders are used afterwards. This technique permits the jeweler to solder a piece together without melting previously made joints.

Soldering: The process of uniting two pieces of heated metal together with fusible flux agents (solder). The two metal pieces remain solid and do not reach the melting point.

Welding: The process of fusing two pieces of the same alloy by heating and allowing the metals to flow together at their juncture without solder.

Work hardening: The result of cold-working a piece of metal by processes such as hammering, squeezing, bending and drawing during which the grain structure becomes compacted. The metal thus becomes less malleable and tougher. Even though the chemical composition does not change, the metal becomes harder. For example, the Vickers hardness of annealed silver is 22, but it is 100 after being cold worked.

2

Gold

A Brief History of Gold

In the Bible, gold is both the first metal and first atomic element to be mentioned. Adam and Eve didn't have far to go to find gold.

> And a river went out of Eden to water the garden; and from thence it was parted, and became into four heads. The name of the first is Pisin: that is it which compasseth the whole land of Havilah, where there is gold; And the gold of that land is good:
>
> Genesis 2:10-12.

Primitive man was able to find gold in its free form in riverbeds when obtaining water. As a nugget, gold could be an attractive jewelry item. Man soon linked gold to the sun god, who provided him with light, warmth and crops. Consequently, gold became important for the religious objects and ceremonies of early civilizations.

Around 700 BC, gold became a basis for economic life. Gyges of Lydia (what is now south-central Turkey) established the first mint to put the seal of his kingdom on uniform lumps of gold. (These gold lumps were an alloy of about 75% gold and 25% silver called "electrum.")

About 550 BC, the Lydian king, Croesus, became the first to introduce coins of pure gold. His royal stamp on the coin was a guarantee as to their weight and purity. Since then, gold coins have been minted by almost every government. The circulation of these coins with the seal of the government served as an ideal advertisement for gold. "Gold" and "high value" became synonymous.

Fig. 2.1 Granulated gold earring (4th to 7th century AD). *Jewelry and photo from Adin Fine Antique Jewellery in Antwerp.*

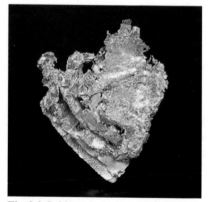

Fig. 2.2 Gold specimen from Breckenridge, Colorado. *Gold from Pala International; photo by John McLean.*

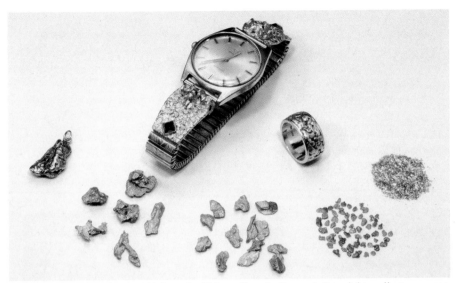

Fig. 2.3 Nugget jewelry & gold from the Yukon, Canada. Lower left to right: collector nugget, large nuggets, small nuggets, coarse gold and gold dust. *Gold and photo courtesy William Newton.*

It was primarily the lure of gold that brought so many Spaniards to Mexico and South America during the 16th century. Three hundred years later, in 1849, one of the greatest gold rushes of all time drew more than 40,000 diggers to California. The population of Australia grew from around 400,000 in 1851 to 1,200,000 in just ten years largely as the result of the discovery of gold. In 1896, more than 60,000 men flocked to the Yukon in Canada near Alaska. Gold panning continues there still. Gold in Brazil, British Columbia, South Dakota, Nevada, Colorado, North Carolina, Georgia, and New Zealand have also created mass movements of population.

The largest gold find ever was in 1886 at the Witwatersrand Reef in South Africa. Since then, South Africa has been one of the world's largest producers of gold along with China, Australia, Peru, the U.S., Russia and Canada,   The gold rushes of the 19th century had a phenomenal impact. They stimulated shipping, commerce and manufacturing throughout the world. The huge increase in the supply of gold inflated the world's currencies and led to the adoption of the gold standard by most of the leading nations (although later it was abandoned).

The transportation system of the United States was significantly expanded in order to deliver supplies to miners throughout the West. Australia and the western part of North America became important commercial centers. A large percentage of the people in these areas would not be there today had it not been for the quest for gold.

On a worldwide and historical basis, gold has always remained the king of money. It's accepted everywhere as a medium of exchange. This, combined with its portability, has made it the most important investment for people living under oppressive or unstable governments. Many Vietnamese refugees, for example, were able to start a new life in other countries, thanks to gold.

Gold has also played an important role as an expression of affection. With the exception of Japan, where brides prefer platinum wedding rings, a gold band has been the traditional choice for wedding rings in most other cultures. Gold is always a welcome gift.

Gilded and Gold Rings Through the Ages

Fig. 2.4 Renaissance fire gilded silver ring with clasped hands, a common form of marriage ring and occasionally friendship ring; circa 1650.

Fig. 2.5 Poesy 22K gold ring with the engraving, "Time Shall Tell I Love You W.ell." Date: late 17th to 18th century.

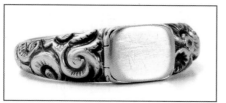

Fig. 2.6 Georgian locket ring in 14K yellow gold, which once contained a locket of plaited hair; circa 1820.

Fig. 2.7 Victorian nine-sided 10K rose gold band engraved with the design of birch leaf; circa 1890.

Fig. 2.8 Edwardian diamond buckle ring of 18K yellow gold; circa 1914; English in origin.

Fig. 2.9 Art Deco buckle and strap motif ring of 9K gold; circa 1930.

Fig. 2.10 Retro palladium and 14K rose and yellow gold eternity band. Retro jewelry is often multi-colored; circa 1940.

Fig. 2.11 Retro 14K white gold wedding ring eternity band; circa 1940. The fluted pie crust perimeter is a retro-era design element.

All rings and photos on this page are courtesy Lisa Stockhammer-Mial, collector, historian and owner of **The Three Graces Antique Jewelry**.

Karat Value (Gold Fineness)

Legally, gold is considered to be pure gold or 24K if it is at least 99% gold. Pure gold is **alloyed** (combined) with other metals to make it more durable and affordable or to change its color. The purity of alloyed gold is described by its fineness, parts per 1000, or by the **karat**, a 1/24 part of pure gold by weight. (The spelling "carat" is used in the British Commonwealth). Every country designates specific qualities as their legal gold standards. The table below lists karat qualities of gold jewelry.

Table 2.1 Gold content and notation

Karats (USA)	Parts Gold	Gold %	Fineness	Notes
24K	24/24	99.9%	999 or 1000	Pure gold
24K		99.0%	990	Minimum allowed for pure gold jewelry; popular in China
22K	22/24	91.6%	916 or 917	Popular in India
21K	21/24	87.5%	875	Popular in the Middle East
19.2	19.2/24	80.0%	800	Standard in Portugal
18K	18/24	75.0%	750	Standard intl. karatage for high quality jewelry; minimum standard in Saudi Arabia,
15K	15/24	62.5%	625	Used in Great Britain from 1854-1932.
14K	14/24	58.3%	585	583 or 58.3% in U.S., minimum standard in Austria, Kingdom of Bahrain, Thailand,
12K	12/24	50.0%	500	Used in Great Britain from 1854-1932.
10K	10/24	41.6%	416 or 417	Minimum standard in the U.S.
9K	9/24	37.5%	375	Minimum standard in Australia, Canada, Israel, Italy, New Zealand, the U.K. and for the Hallmarking Convention.
8K	8/24	33.3%	333	Minimum fineness in Denmark

Sources: www.gold.org/jewellery/about_gold_jewellery/caratage/, *CIBJO Precious Metals Book*, www.ftc.gov/bcp/guides/jewel-gd.shtm; *World Hallmarks, Volume 1, 2ⁿᵈ Edition* by William B. Whetstone, Danusia V. Niklewicz, & Lindy L Matula; GIA *Gold & Precious Metals Course*, www.bis.gov.uk/assets/britishhallmarkingcouncil/docs/hallmark-guidance-notes-final.pdf.

375   585   750   916   990   999

Fig. 2.12 Metal and fineness marks for 9K, 14K, 18K, 22K & 24K gold in the U.K. The metal type is indicated by the shape of the surround.

Fig. 2.13 Dutch yellow and red 14K gold filigree dangle earrings, circa 1880's. *Jewelry & photo from Adin Fine Antique Jewellery in Antwerp, Belgium.*

Fig. 2.14 Textured 18K gold bracelets by Katy Briscoe. *Photo courtesy Katy Briscoe.*

Fig. 2.15 A 9-carat hallmark and English maker's mark of H&Co. *Cufflinks and photo courtesy The Three Graces Antique Jewelry.*

Fig. 2.16 The French eagle head mark for 18K gold. *Jewelry and photo courtesy The Three Graces Antique Jewelry.*

Fig. 2.17 Horse head mark used in France for gold between 1847–1919 to indicate the item was assayed outside of Paris. *Jewelry & photo: The Three Graces Antique Jewelry.*

Fig. 2.18 Top: Italian made 14K chain, bottom: lighter yellow 14K American-made chain. *Photo © by Renée Newman.*

Which is Better—14K or 18K Gold?

You may wonder whether 14K or 18K gold is better, particularly when selecting an everyday ring. 18K rings have 3/4 gold and 1/4 other material whereas 14K is just a little more than half gold. Consequently, rings with 18K gold are more valuable. They're also less likely to cause a reaction in people who are allergic to metals alloyed with gold, and they usually have a deeper yellow color than 14K gold. However, some 18K and higher karat gold alloys may not be as hard and strong as 14K gold.

Rings of 14K gold are less expensive and often wear better. In North America, you'll probably have a better selection of 14K jewelry because a greater variety of it is manufactured. However, 14K might have a tendency to discolor or tarnish due to the lower percentage of gold and high percentage of copper. Occasionally, the metals alloyed with 14K gold cause an allergic reaction in some people. Nickel white gold alloys are the ones most likely to cause allergic reactions. (White gold is made by alloying pure gold with metals such as palladium, silver, zinc and/or nickel. Many European countries have laws forbidding the use of nickel in jewelry.)

Much of the better jewelry is made in 18K gold, palladium or platinum. More and more jewelry of 22K, 24K or 990 gold is becoming available. This higher-karat gold is hypoallergenic and resists tarnish, and some of the new alloys are relatively durable. If you have a ring custom made, you can choose the gold percentage.

Have a look at some 14K, 18K and higher-karat gold jewelry. Usually there's some difference in color. Consider your color preferences along with the above points when choosing the karat quality. More often than not, the determining factor will be whether or not you can find jewelry you like in your price range.

Gold Colors and Alloys

Gold is generally described as yellow, but it doesn't have a true yellow color. To prove this, find an object that looks pure bright yellow and compare it to a piece of gold jewelry. The colors will be different.

Because of gold's metallic luster and the way in which it reflects its surroundings, it's not easy to determine its color. However, when you analyze it carefully, you discover that the color of gold is a combination of light brown, yellow and orange. Nevertheless, gold is referred to as a yellow metal.

If you've ever compared a 14K Italian-made chain to one made in the United States, you've probably noted that the American-made one has a lighter color. This is normally because it contains more silver and less copper than the one made in Italy. Sometimes, too, Italian 14K chains are plated with 18K or 24K gold to produce this effect. Europeans generally prefer a deeper gold color and so do Asians. Another reason why gold is sometimes plated is to make the color more uniform.

Even within Europe, yellow gold will vary in color; consequently it may be hard to match gold mountings and their components (clasps, settings, etc.). To alleviate this problem, gold sample plates for standard alloys are used there.

Gold alloys not only come in varying shades of yellow, they may also be green, pink, or white. The pinker the alloy, the more copper it contains. The greener it is, the more silver it has.

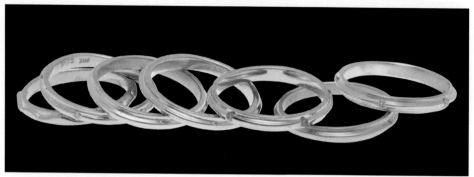

Fig. 2.19 Stacking rings with 14K and 18K gold colors that designer Eve Alfillé describes from left to right as blush, blond, green, blond, blush, rose and white. *Photo by Matthew Arden.*

Fig. 2.20 An 18 carat rose gold ring engraved with a Moors head seal. *Ring and photo courtesy R.H. Wilkins Engravers, Ltd in London.*

Fig. 2.21 A hand made, hammer chased, and fluted 18 carat gold cuff by designer goldsmith Wayne Victor Meeten. *Photo: Gavin Cottrell.*

Fig. 2.22 Pure 24K gold drop earrings from Gurhan's Splash collection. *Photo by Gurhan.*

Fig. 2.23 Yossi Harari 24K gold & oxidized Gilver™ cuff. *Photo courtesy Muse Imports.*

There are various white gold alloys but the most common ones in the U.S. contain gold with nickel and/or palladium, and may include silver, zinc, copper and other metals. Most of these white gold alloys were developed after World War II as a substitute for platinum. Prior to the war, platinum was the preferred choice for diamond rings in America, but when the U.S. government declared platinum a strategic metal, it was no longer available for jewelry use. By the early 1960's, yellow gold had become the preferred metal for diamond rings, and white gold tended to be associated with silver, not platinum. This trend has reversed now that platinum has become so popular. Palladium and silver are also being used for diamond rings along with various white gold alloys

Gold alloys that look blue, black or purple do exist, but they're not often made into jewelry because they're extremely difficult to work with. Curiously, when purple gold is formed by melting gold and aluminum together, the resulting compound loses some of its metallic properties.

A table indicating the composition of some different 18K gold alloys is presented below. It will give you a better understanding of how colored gold is formed. Other formulas are also used to create various gold colors.

Table 2.2 Some 18K Gold Alloys and Their Colors

%	gold	silver	copper	palladium	zinc	nickel	iron	cadmium
light yellow	75	16	9					
yellow	75	12.5	12.5					
green	75	25						
med. green	75	5	20					
red/rose	75		25					
blue	75						25	
grey	75		8				17	
pink/rose	75	5	20					
pink	75	9	16					
brown	75	6.25						18.75
white	75			25				
white	75		10		1	14		
white	75	10.5	3.5	10		1		

Sources: *The Goldsmith,* Dec. 1984, *Professional Goldsmithing* by Alan Revere, *Fundamentals of Metalsmithing* by Tim McCreight.

Occasionally surface techniques are used to color karat gold. In the chapters on gold testing, it is mentioned that nitric acid will turn 10K gold dark brown. Gold can also be darkened through oxidation when heated.

Fig. 2.24 Note the natural looking coloring of the migrant fieldfare (thrush) of this 9K gold brooch. The orange, brown, and black colors were created by heating the gold till it turned black through oxidation. Later the craftsman artfully worked down the metal to expose the gradations of color. *Brooch by Bert McCrum; photo by the designer, Alan Hodgkinson.*

Figure 2.24 shows how a naturalistic coloring was achieved on a bird brooch by first oxidizing the 9K gold to a black color. Later the other colors were exposed using scraping techniques.

Gold can also be plated to change its color. Rhodium plating is commonly used to make pale yellow gold alloys look white. See Chapter 12 for more information on plating, as well as coating and enameling.

In 2012, the BBC and the *International Online Journal of Optics* reported that scientists in Southampton, U.K. are now able to change the color of gold by embossing tiny raised or indented patterns on the metal's surface, which alters the way it absorbs or reflects light. However, as with platings, coatings and oxidizing methods, embossing is not a permanent coloring technique.

3

Platinum

Platinum (Pt) is the most famous of the platinum group metals. Therefore it's not surprising that it lends its name to the family of six metals which have similar chemical and physical characteristics—palladium, iridium, platinum, osmium, rhodium and ruthenium. These metals, which are often deposited together, are commonly called the **platinum group metals** or **PGM**. The metals are neighbors in the periodic chart of chemical elements, which explains why they are similar.

The history of platinum dates back more than three thousand years, beginning with the ancient civilization of Egypt. Archaeologists have found Egyptian gold pieces from 1400 BC that contain traces of platinum. They also uncovered a gold and silver box with a tiny platinum panel dating back to the seventh century BC.

The first people who may have known that platinum was something other than silver or gold were native South Americans in the area that is now Colombia. They developed a crudely refined form of platinum about 1500 years before Europeans discovered it. The Spanish conquistadors were unable to appreciate the platinum they found in South America, however. In fact, they considered it a nuisance because they didn't know how to melt or work it, and they derogatorily named it *platina* (little silver).

Platinum came to the attention of European scientists in the mid-1700's. Several people experimented with it, but its high melting point made progress difficult. It wasn't until the early 1800's that Englishmen William Hyde Wollastan and Smithson Tennant produced the first commercial-grade platinum. After new deposits of platinum were found in Russia in 1822, it began to appear in decorative chains. By the 1850's, it was featured in cuff links and shirt studs. Yet platinum remained fairly obscure until the 1890's, when French jeweler Louis Cartier began using it. Since Cartier's clientele included most of the crowned heads of Europe, platinum soon became one of the jewelry world's greatest status symbols. In 1908, the 516-carat Star of Africa diamond was mounted in platinum on the British royal scepter. Platinum became even more popular in the 1920's and '30's, especially for Art Deco jewelry. It was also the preferred metal for engagement and wedding rings. Even though it could display a delicate beauty, platinum was sturdy, it resisted wear, and it held stones securely.

Platinum's flourishing reign in jewelry came to an abrupt halt with the outbreak of World War II. The U.S. government declared it a strategic metal and banned its use in all nonmilitary applications, including jewelry. Because of strong consumer preference for platinum's neutral color, white gold was substituted for platinum (white gold's color is created by adding other metals to yellow gold). Currently, the two most commonly used white gold alloys in the US are nickel and palladium. Each individual alloy manufacturer has their own mixture which may include silver, zinc, copper and other metals in addition to either nickel or palladium. After the embargo was lifted, platinum did not regain its previous popularity because white gold was less expensive and easier to work with and the public accepted the change of metal.

It wasn't until the early 1990's that platinum began to reemerge as an important jewelry metal in North America. Japanese consumers, however, were already aware of its advantages. About 90% of all wedding and engagement rings sold in that country were being made of platinum. Platinum had not been banned in jewelry during World War II in Japan. Instead, the Japanese government had considered gold to be a strategic metal.

In 1992, Platinum Guild International USA (PGI) was formed for the purpose of reeducating American consumers about the durability and elegance of platinum. Technical training was also offered to jewelers to help them deal with the challenges of working with platinum. PGI's efforts have paid off. Platinum jewelry is now widely available and large numbers of Americans are enjoying its benefits. Its popularity has even spread to China and India. Platinum has regained its former preeminence in the jewelry market.

Platinum Group Metals

English chemist William Woolaston isolated **palladium** in 1803 while trying to find a way to produce malleable platinum for commercial use. He named it after the planet Pallas, which had also just been discovered. Palladium is discussed separately in Chapter 4 because it has become an important jewelry metal in its own right.

After discovering palladium, Woolaston isolated rhodium. This metal derived its name from the Greek *rhodon* for rose because rhodium compounds formed during the refining process are rose-colored.

Woolaston's partner, Smithson Tennant, discovered osmium and iridium. Later a Russian scientist named Claus discovered ruthenium, whose name originates from the Latin word for Russia *Ruthina.*

Until 1916, the two main sources of supply of platinum group metals were South America and Russia. Today the largest producer is South Africa. Russia is in second place, followed by Canada, the United States and Zimbabwe.

Iridium (Ir) (5% to 15%) is commonly added to platinum to make it harder and more suitable for jewelry wear. Platinum consisting of 10% iridium and 90% platinum is a popular alloy in the U.S. Iridium is also used as an industrial plating material.

Ruthenium (Ru) (5%) is frequently used to increase platinum's hardness (resistance to scratching). 950 Pt 50 Ru serves as a general purpose alloy in Europe, Hong Kong and the United States. Tiffany & Co. carries scratch-resistant ballpoint pens which are made of ruthenium. They're ideal for people who are rough on their pens. Ruthenium is also used for plating silver and other metals.

Rhodium (Rh) is a popular metal for plating. Silver jewelry is often rhodium-plated to prevent it from tarnishing. White gold settings are commonly rhodium-plated to harden the surface and to increase the contrast between yellow gold and white gold components of the same jewelry piece. Since no white-gold alloy is truly white, most finished white-gold jewelry is rhodium-plated to increase the intensity of the white color as well as to improve the hardness of the surface. However, if a white-gold piece is rhodium-plated, it's more difficult to repair or resize because the rhodium layer will burn in the heat of the soldering torch, flaking off the jewelry piece and making it impossible for the solder to flow. Therefore before any heating takes place, the rhodium has to be removed from the surface. This is costly and time consuming.

Fig. 3.1 A rare intergrown cluster of platinum crystals from Konder, Russia with gold deposition on the surface. *Specimen from Rob Lavinsky of the Arkenstone; photo by Joe Budd.*

Fig. 3.2 Platinum (950) chain hand fabricated by Helene M. Abitzsch. *Photo courtesy of the Platinum Guild International (Germany).*

Fig. 3.3 Platinum ring hand forged by Stephen Vincent. *Photo by Louisa Marion Photography.*

Fig. 3.4 Platinum diamond ring design © by Eve J. Alfillé. *Photo by Matthew Arden.*

Fig. 3.5 Platinum diamond ring and photo from Mark Schneider Design.

Fig. 3.6 Platinum diamond ring and photo from Mark Schneider Design.

Another disadvantage of rhodium plating is that the rhodium layer wears off over time and has to be replated. Consequently, white-gold pieces, especially rings, are considered to be high maintenance.

Platinum, by comparison, is white throughout and doesn't require plating. If a jeweler finds it necessary to rhodium-plate a platinum piece, this usually indicates there's a manufacturing problem that needs to be hidden from the customer. It could be that different platinum alloys (with different metal colors) were used in the assembly of the piece and that plating is required to provide a uniform color. Perhaps the jeweler did not weld the piece but used platinum solder, which contains little or no platinum and has a different color than the body color of the platinum jewelry. In order to mask the darker color of the soldering seam, the piece is exposed to a uniform color-providing plating process. (In 1999, some new 90% to 95% platinum solders were developed by Precious Metals West in Los Angeles to help alleviate platinum solder problems.)

Osmium is the hardest metal known and is not used much in jewelry manufacturing. It's extremely rare and difficult to work with. Osmium wire has been used in light bulbs because it can withstand repeated expansion and contraction (on/off) without breaking.

Platinum Fineness Regulations Worldwide

The purity of platinum is described only in terms of fineness throughout the world. However, the markings and alloys used may vary slightly. **PT**, the chemical symbol for platinum, and **PLAT** are used as abbreviated marks for platinum when identifying the metal and its fineness. For example, when PT900 or PLAT900 is stamped on a jewelry piece, it means that it contains at least 900 parts per thousand of pure platinum. The unmodified terms PLATINUM, PLAT and PT generally refer to an alloy containing at least 95% platinum, in other words 950 parts per 1000 pure platinum.

In most countries of the world, the minimum fineness standard required for platinum is 850 parts per 1000 pure platinum. The United States is an exception. In 2010, the Federal Trade Commission (FTC) Guidelines were amended to designate alloys with only 50% platinum (and the balance with base metals such as copper and cobalt) as a platinum alloy; the FTC guidelines are fully subject to the National Gold and Silver Stamping Act enacted in 1905. Violations of FTC standards can lead to an FTC civil enforcement action that can impose both fines and injunctive relief. Details of the revised FTC Platinum Guidelines are shown in the yellow text box. Even though alloys with 50% platinum can be considered platinum in the U.S., it's advisable for manufacturers who wish to sell jewelry outside the US to use alloys containing 95% platinum. Otherwise, they won't be able to legally sell it as platinum jewelry in countries such as Canada, which has a minimum platinum fineness standard 950 parts per 1000 pure platinum. More information on the Canadian Standard from the Guide to the Precious Metals Marking Act and Regulations Enforcement Guidelines (July 4, 2006) is found at www.competitionbureau.gc.ca/eic/site/cb-bc.nsf/eng/01234.html.

In the UK and most of Europe, the minimum fineness standard is 850 parts per 1000 pure platinum. However, PLAT 850 is most commonly used for chain whereas PLAT 950 is typically used for ring mountings. PLAT 900 is also an accepted standard for many countries. For example, the United Kingdom allows four fineness standards: 999, 950, 900 and 850

Table 3.1 Platinum content and notation

Name	Platinum %	Fineness Stamp	Notes
Fine platinum	99.9%	999 or 1000	Pure platinum, most common purity for Japanese wedding bands
PLATINUM, PLAT, PT	95%	950	Most common platinum alloy fineness standard throughout the world; minimum standard in Canada, Kingdom of Bahrain, New Zealand, The Netherlands, Poland, Romania. In India, the only purity receiving a certificate from Platinum Guild Intl.
	90%	900	Minimum standard in Hungary, India and Thailand; most common purity in China; most common purity for Japanese jewelry
	85%	850	Minimum standard in Australia, China, Hong, Kong, Israel, Japan, Saudi Arabia, the UK and most European countries.
	80%	800	A permissible purity in the US
	50%	500	Lowest platinum standard and the minimum fineness standard in Bulgaria, Lithuania and the U.S. Any alloy with less than 50% cannot be called platinum in countries that have established a platinum standard.

Sources for table and preceding text include: www.cibjo.org; *CIBJO Precious Metals Book*, www.ftc.gov/bcp/guides/jewel-gd.shtm; *World Hallmarks, Volume 1, 2nd Edition* by William B. Whetstone, Danusia V. Niklewicz, & Lindy L Matula; GIA *Gold & Precious Metals Course,* Assignment 6, *Platinum Starter Kit Bench Companion* by the Platinum Guild International, www.bis.gov.uk/assets/britishhallmarkingcouncil/docs/hallmarking-act-1973-with-palladium.pdf, www.platinumguild.com/output/page1687.asp, www.competitionbureau.gc.ca.

When buying platinum jewelry, make sure it has a fineness mark or is stamped "PLATINUM," "PLAT" or "PT." Some countries, such as Japan, place the responsibility on the public not to buy unmarked jewelry. If you're buying platinum jewelry abroad, verify that it conforms to the standards of your country of residence if there's a possibility you may wish to resell it or pawn it. If the jewelry has a purity of at least 95%, you needn't be concerned. You can resell 950 PT jewelry anywhere in the world.

  Metal and fineness marks for platinum in the U.K. The metal type is indicated by the shape of the surround.

For additional information on fineness marking, go to:
www.bis.gov.uk/britishhallmarkingcouncil, British Hallmarking Council
www.hallmarkingconvention.org, the Hallmarking Convention
www.theiaao.com, the International Association of Assay Offices
www.jvclegal.org/JVC_Platinum_Guide.pdf, Jewelers Vigilance Committee

FTC Platinum Guidelines

from www.ftc.gov/bcp/guides/jewel-gd.shtm

§23.7

(b) The following are examples of markings or descriptions that may be misleading:

(1) Use of the word "Platinum" or any abbreviation, without qualification, to describe all or part of an industry product that is not composed throughout of 950 parts per thousand pure Platinum.

(2) Use of the word "Platinum" or any abbreviation accompanied by a number indicating the parts per thousand of pure Platinum contained in the product without mention of the number of parts per thousand of other PGM contained in the product, to describe all or part of an industry product that is not composed throughout of at least 850 parts per thousand pure platinum, for example,"600Plat."

(3) Use of the word "Platinum" or any abbreviation thereof, to mark or describe any product that is not composed throughout of at least 500 parts per thousand pure Platinum.

(c) The following are examples of markings and descriptions that are not considered unfair or deceptive:

(1) The following abbreviations for each of the PGM may be used for quality marks on articles: "Plat." or "Pt." for Platinum; "Irid." or "Ir." for Iridium; "Pall." or "Pd." for Palladium; "Ruth." or "Ru." for Ruthenium; "Rhod." or "Rh." for Rhodium; and "Osmi." or "Os." for Osmium.

(2) An industry product consisting of at least 950 parts per thousand pure Platinum may be marked or described as "Platinum."

(3) An industry product consisting of 850 parts per thousand pure Platinum, 900 parts per thousand pure Platinum, or 950 parts per thousand pure Platinum may be marked "Platinum," provided that the Platinum marking is preceded by a number indicating the amount in parts per thousand of pure Platinum (for industry products consisting of 950 parts per thousand pure Platinum, the marking described in § 23.7(b)(2) above is also appropriate). Thus, the following markings may be used: "950Pt.," "950Plat.," "900Pt.," "900Plat.," "850Pt.," or "850Plat."

(4) An industry product consisting of at least 950 parts per thousand PGM, and of at least 500 parts per thousand pure Platinum, may be marked "Platinum," provided that the mark of each PGM constituent is preceded by a number indicating the amount in parts per thousand of each PGM, as for example, "600Pt.350Ir.," "600Plat.350Irid.," or "550Pt.350Pd.50Ir.," "550Plat.350Pall.50Irid."

Platinum Alloys

Platinum jewelry can differ not only in the amount of pure platinum present but also in the metals that are combined with the platinum. Iridium is probably the most widely used metal in platinum alloys, but cobalt, copper, palladium and ruthenium are also popular alloying ingredients. The percentage and type of metals combined with the platinum affect the alloy's physical properties such as hardness, density, strength, color and melting temperature.

Fig. 3.7 Fig. 3.2 Platinum (950) earring hand fabricated by Helene M. Abitzsch. *Photo from Platinum Guild International (Germany).*

Fig. 3.8 An 18K rose gold and platinum bracelet design by Michael Zobel featuring an 86.38-ct aquamarine (Becker). *Photo courtesy Atelier Zobel: photo by Fred Thomas.*

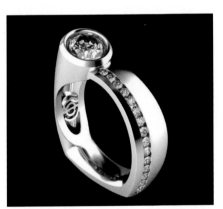

Fig. 3.9 Platinum and diamond ring. *Jewelry and photo by Erik Stewart.*

Fig. 3.10 Platinum and diamond ring. *Jewelry and photo by Erik Stewart.*

Fig. 3.11 Platinum & 18K gold diamond ring. *Jewelry and photo from Varna Platinum.*

Fig. 3.12 Platinum and diamond ring. *Jewelry and photo from Varna Platinum.*

Some of the most common platinum alloys are described below.

Pt 5Ir (95% platinum 5% iridium) - This alloy has a low hardness (Vickers 80) but a high work-hardenability. It is recommended for welding but not for casting. The hardness can be increased by replacing a portion of the iridium with gallium.

Pt 10Ir (90% platinum 10% iridium) - Hard and strong and easy to work and weld, this is the most popular general purpose alloy in the United States. It's also used in Germany and Japan.

Pt 5Co (95% platinum 5% cobalt) - recommended for casting, this alloy is widely used in Europe, Hong Kong and the United States. Its fluidity (viscosity) allows for extremely fine castings of filigree and thin-cross-section pieces. The mechanical properties for fabricating and workbench repairs are good; however, this alloy has a tendency to oxidize when exposed to the jeweler's torch. Pt5Co has a Vickers hardness of 135. Cobalt platinum alloys are slightly magnetic, which is regarded by some people as an undesirable attribute for a precious metal. In fact, there have been cases where jewelry professionals have misidentified Pt5Co as a nonprecious metal because it was slightly attracted to a magnet.

Pt 5Ru (95% platinum 5% ruthenium) - This is a hard (Vickers 130), ductile, general purpose alloy. Advances in equipment and technology, along with the talents of experienced, professional casting technicians, have made Pt 5Ru a desirable alloy which is now used all over the world. Tiffany & Co. mastered the use of Pt5 Ru in platinum casting because it didn't want to offer slightly magnetic Pt 5Co merchandise to its customers. Magnetism in jewelry metals is usually considered to be a characteristic of less valuable metals such as iron and steel. If jewelry is fake, it can often be detected with a magnet.

Pt 5Pd (95% platinum 5% palladium) - Used in Japan, Europe and Hong Kong, Pt 5Pd is recommended for delicate settings and die-stamped jewelry. It has a low hardness (Vickers 60–68), lower than Pt 5Ir, Pt 5Co, Pt 5Ru and Pt 5Cu.

Pt 10Pd (90% platinum 10% palladium) & **Pt 15Pd** (85% platinum 15% palladium) - These alloys are used in Japan and Hong Kong. They have a dull gray color and a low hardness, which prevents them from having a good shiny finish. Consequently, these alloys are usually rhodium plated to give them a brighter finish and a harder surface. Pt 15Pd is used widely in Japan for chain making and cast jewelry.

Pt 5Cu (95% platinum 5% copper) - This is a good general purpose alloy which is popular in Europe, especially in Germany. Platinum jewelry in Hong Kong is also made with this alloy. The fact that Pt 5Cu is alloyed with copper prevents it from being as white as other platinum alloys. It has a Vickers hardness of 120 in the annealed state and 108 when cast.

Most of the preceding information on platinum alloys is from a report entitled "Platinum Alloys and Their Application in Jewelry Making" by the Platinum Guild International USA. It was written by Jurgen Maerz, Director of Technical Education, with assistance from Taryn Biggs & Stefanie Taylor of Mintek, South Africa, Johnson Matthey, Engelhard-Clal, Imperial Smelting, C. Hafner Co. and Techform LLC. For additional platinum information see:

www.preciousplatinum.com, www.preciousplatinum.co.uk & www.platinumguild.com

4

Palladium

Palladium (Pd), a member of the platinum group metals, offers many of the advantages of platinum but weighs only 56% as much, making large pieces more comfortable to wear, especially earrings and necklaces. (The specific gravity of palladium is 12.2 compared to 21.4 of platinum.) The lower density and weight of palladium also allow for chunkier and taller designs that are still affordable.

Palladium pieces are also more affordable because the price of palladium has usually been significantly lower than that of platinum. In 2012, palladium sold for around $575–$700/oz, or generally less than half the price of gold and platinum, whereas gold typically ranged from about $1550–$1775/oz and platinum from around $1400–$1700/oz.

A Brief History of Palladium

Considering its status as a precious metal and member of the platinum group, you may wonder why palladium has only been promoted as an ideal jewelry metal since about 2006. It wasn't until 1803 that palladium was isolated from platinum and identified as a separate elemental metal by a researcher named William Hyde Wollaston. He named his discovery after the asteroid Pallas, which was in turn named after the Greek Goddess of Wisdom.

Palladium, however, was quite rare until 1930 when the International Nickel Company of Canada began producing it in significant quantities. In 1931, a German company named Heraeus developed and patented alloys of palladium with silver and gold. These alloys were ideal for dentistry, and are still used today in bridges and crowns. Jewelry designers started working with palladium in the late 1930's. During World War II, it was used as an alternative to platinum, which was declared a strategic metal. Starting in the 1970's, palladium was in demand for catalytic converters and for electronic circuitry, but because of its low supply and high price relative to other metals, it was seldom promoted as a jewelry metal. In early 2001, the price of palladium reached a high of $1090 per ounce, higher than the price of platinum at the time. Russia had stopped selling palladium, and car manufacturers needed it. Like all metals, the price of palladium will fluctuate.

The majority of the world's palladium has come from Russia and South Africa. Fortunately there have also been finds of palladium in Montana, Canada and Zimbabwe, which have helped increase the supply and bring down prices below those of gold and platinum. As a result, it makes sense to take advantage of palladium's benefits and promote it as a distinctive jewelry metal instead of using it simply as a component of white gold for jewelry.

Fig. 4.1 Palladium and Tahitian pearl pendant and ring by Alishan. *Photo by John Parrish.*

Fig. 4.2 Palladium & sapphire ring. *Ring and photo from Stephen Vincent Design.*

Fig. 4.3 Palladium and 18K red gold band. *Ring and photo from Stephen Vincent Design.*

Fig. 4.4 Palladium and diamond ring. *Jewelry and photo from Varna.*

Fig. 4.5 Palladium and diamond ring. *Jewelry and photo by Erik Stewart.*

Fig. 4.6 Palladium and diamond ring. *Jewelry and photo by Erik Stewart.*

Fig. 4.7 Palladium signet ring and wax impression. *Ring and photo from R. H. Wilkins Engravers Ltd.*

Fig. 4.8 Palladium and mabe pearl pendant by Alexandra Hart. *Photo: Robert Weldon.*

Fig. 4.9 Palladium/14kw black diamond ring & photo by Sara Commers of Studio C Designs.

Fig. 4.10 Palladium and black diamond bracelet by Todd Reed. *Photo by Brian Mark.*

Fig. 4.11 The "9ct & PALL" stamp indicates this locket (circa 1920) is made of 9K gold and palladium. *Locket & photo from The Three Graces Antique Jewelry.*

Fig. 4.12 Palladium &18K gold spessartine ring. *Jewelry and photo by Erik Stewart.*

In March 2006, an organization called the Palladium Alliance International (PAI) was formed to position palladium as a luxurious jewelry metal. PAI provides education, marketing and technical support to the trade and general public. It has offices in Billings, Montana, and Shanghai, China. PAI's website is www.luxurypalladium.com.

Palladium and platinum are harder than silver and gold. However, the actual hardness and durability of jewelry metals depends on the alloy used as well as the manufacturing process. For example, machine-made (die-struck) settings are usually harder than those that are cast.

Some palladium alloys offer the same high 95% purity as platinum, while other palladium alloys may be as low as 50% or 500 parts per 1000. Palladium may be alloyed with ruthenium, iridium, copper, gold, silver, cobalt and/or other metals. According to the *Palladium Technical Manual, UK Edition*, three important palladium 950 alloys are:

95% palladium – 5% ruthenium: commonly used for wedding bands. Ruthenium makes palladium harder and stronger than pure palladium.

95% palladium – 5% ruthenium and gallium: used for casting, this alloy is hard and resists wear.

95% palladium – 5% copper: used by Italian companies for chain making

Like platinum, palladium has the following advantages. Palladium:
♦ Resists tarnish, doesn't discolor the skin or turn yellowish with wear
♦ Is malleable, making it easy to form and manipulate. It doesn't have the brittleness of white gold.
♦ Is hypoallergenic
♦ Is whiter than white gold. Palladium is about the same color as platinum but slightly darker and more grayish. The color depends on the alloy used. When consumers see palladium next to platinum jewelry, they usually think it looks the same. Unlike gold, palladium doesn't need to be repeatedly rhodium plated to maintain its appearance as a white metal.
♦ Is classified as a precious platinum group metal with an inherent commodity value and currency code
♦ Is strong and wears well
♦ Palladium 950 provides a higher purity than white gold, while at the same time having a whiter color.

Working with palladium requires training and different tools than those used for gold and silver. For example, the Palladium Alliance International says that casting equipment with a protected atmosphere is required for both the melting and casting parts of the casting operation to prevent palladium from absorbing gas and developing porosity. Welding palladium in a laser welder requires an inert gas cover. Best results are obtained with lower power settings, longer delays and medical grade argon as the cover gas. However, at the bench, palladium must be soldered. The Palladium Alliance International offers free seminars, videos, information and support for anyone interested in learning how to work with palladium.

Palladium Fineness Standards

The three most common palladium fineness standards in the world are 999, 950 and 500. The United States has not yet enacted a law regarding palladium standards, but the British Parliament did in 2010. As of January 2010, all palladium articles in excess of one gram must be hallmarked. The three British fineness standards are 999, 950 and 500 and the solder fineness must be at least 700 parts per thousand precious metal content for 950 and 500 fineness pieces.

Www.hallmarkingconvention.org, (the website of the Convention on the Control of Articles of Precious Metals) lists the palladium fineness standards of its members and indicates that an 850 fineness is allowed in Finland, Lithuania and Poland. CIBJO (The World Jewelry Confederation) upholds the three fineness standards of 999, 950 and 500 and states that the solder for palladium articles of all standards shall contain at least 700 parts per 1,000 of palladium. Table 4.1 summarizes palladium fineness standards.

Table 4.1 Palladium content and notation*

Name	Palladium %	Fineness Stamp	Notes
Fine palladium	99.9%	999	Pure palladium
Palladium	95.0%	950	Most common palladium fineness standard in Europe and North America
	85%	850	A palladium standard found in Finland, Lithuania and Poland
	50%	500	Lowest palladium standard. Any alloy with less than 50% palladium cannot be called palladium in countries that have established a palladium standard.

Sources for table and preceding text include: www.cibjo.org; *CIBJO Precious Metals Book*; *World Hallmarks, Volume 1, 2ⁿᵈ Edition* by William B. Whetstone, Danusia V. Niklewicz, & Lindy L Matula, www.hallmarkingconvention.org, and the *Palladium Technical Manual*, UK Edition.

Left: British palladium 950 metal & fineness mark
Right: International Hallmarking Convention palladium 950 Common Control Mark (CCM). The shape of the diagram indicates the type of precious metal— palladium, in this case.

In the United States, you may find other finenesses besides the ones above. For example, jewelry that has been advertised as **14K white palladium** is 14K gold alloyed with 58.3% palladium, but this is a rare palladium alloy. Sellers are more likely to feature the gold component and sell 14K palladium white gold. This is an alloy of 58.3% gold and 41.7% palladium. A piece made of this alloy may be marked **14K W Pd** in the US. Jewelry made with 95% or more palladium can be stamped **PALLADIUM**, **PD950** or **950PD**. Jewelry with 50% palladium may be stamped **500PD** or **PD500**. Diagrams are not used in the US to identify the metal type.

For additional information on palladium consult:

www.stillwaterpalladium.com

www.luxurypalladium.com

www.palladium.com.cn

www.bis.gov.uk/britishhallmarkingcouncil/new/palladium/

Palladium Technical Manual at www.thegoldsmiths.co.uk

Fig. 4.13 PD stamp on a palladium 950 ring shank.

5

Silver

Many cultures have associated silver with the moon and gold with the sun. Both metals have symbolized wealth and prosperity and are said to intensify the spiritual and healing powers of gems. After the discovery of the Americas by the Europeans, so much silver was mined in Central and South America that the value attached to silver fell sharply. This led to the conversion of most monetary systems to the gold standard.

Today Mexico, Peru, Australia, Russia, Canada and the United States are the leading producers of silver, (chemical symbol "Ag"). Most current silver production stems from recovery of the mining process of lead, copper, gold or zinc.

Prestigious jewelers have offered silver jewelry for decades; nevertheless, before 2008 when gold prices began to skyrocket, silver was generally positioned to compete with fashion jewelry, rather than fine jewelry in gold, platinum and palladium. Today high-end necklaces and diamond rings are often made of silver. A few manufacturers who used to sell only gold jewelry are now selling primarily jewelry made of silver. It is the whitest, brightest and most reflective jewelry metal, it has always been considered precious, and its high malleability and ductility make it ideal for jewelry making.

Pure silver, like pure gold and platinum, is too soft for most jewelry manufacturing techniques so it is often alloyed with other metal(s). Copper is normally preferred because it improves silver's hardness and durability without detracting from the characteristic bright shine of silver. However, even alloyed silver does not hold gemstones as securely as settings made of white gold, platinum or palladium alloys because the tensile strength of sterling silver is lower.

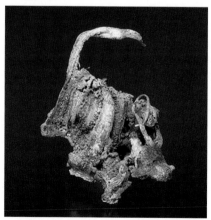

Fig. 5.1 Silver specimen from Freiberg, Germany. *Pala Intl; photo by John McLean.*

Fig. 5.2 Michigan silver specimen & photo: Rob Lavinsky, the Arkenstone.

Fig. 5.3 Sculpture of fine silver by Frank W. Heiser. *Photo from Frank Heiser.*

Fig. 5.4 Recycled sterling silver and wulfenite pin/pendant and photo by Carina Rossner.

Fig. 5.5 Sterling silver earrings design copyright by Eve Alfillé. *Photo by Matthew Arden.*

Fig. 5.6 Sterling silver earrings by So Young Park. *Photo by So Young Park.*

Fig. 5.7 Sterling silver clay jewelry by Holly Gage. *Photo by Holly Gage.*

Fig. 5.8 Sterling silver and carnelian bracelet & photo by Sara Commers/ Studio C Designs.

Sterling silver is the alloy most commonly used in jewelry and usually consists of 92.5% silver and 7.5% copper. Adopted as a standard alloy in England in the 12th century, sterling silver is normally identified with its fineness marking of **925**. Other acceptable markings on sterling are "STER," "STERLING" and "STERLING SILVER."

Britannia silver is defined in the Hallmarking Act 1973 as being of a fineness of 958.4 parts per thousand, while the fineness mark itself is 958 only. Britannia silver replaced sterling silver as the compulsory silver standard in England from 1696 until 1720. Sterling silver was approved again for use by silversmiths in 1720. However, many of the silver bullion coins issued by the Royal Mint since 1998 have been minted in Britannia standard silver. Britannia silver should not be confused with Britannia metal, a pewter-like alloy containing no silver.

Another common silver alloy that contains 10% to 20% copper is called **coin silver**. US dimes, quarters and half dollars minted before 1965 contained 90% silver. An alloy popular in the Far East uses 90% silver and 10% zinc.

In Europe, alloys with 80%, 83%, and 83.5% are also used. Metal with 80% silver is stamped 800 and sometimes called European silver or continental silver. This is probably the second most commonly used silver standard in the world. France and Germany are two important European countries that have used 800 silver.

In Peru, a high percentage of jewelry is made with alloys of 95% silver and stamped 950. Flatware and hollowware in France has been made out of 950 silver, and a few silversmiths in the silver center of Taxco, Mexico have used 950 silver alloys. Occasionally when people see the 950 stamp, they assume that the metal is platinum. Look for "PT" or "PLAT" next to the 950 stamp to determine if it's a platinum stamp.

Silver jewelry sold at street vendors can have varying degrees of fineness, sometimes lower than standard qualities, making it difficult to repair and size the jewelry. Therefore, if you buy rings from tourist vendors, make sure they fit and don't need to be resized.

Besides being a popular jewelry metal, silver is widely used in industry. According to www.silverusersassociation.org, about one-third of the silver produced worldwide is used in photography. Silver is the best conductor of heat and electricity and the most reflective of all elements, making it valuable for solar panels, automobile rear window defoggers, electrical circuits, batteries, computers, pacemakers and space travel. Silver has superior bactericidal qualities; small concentrations of silver or silver salts kill bacteria and the bacteria do not develop resistance to silver, as they do to many antibiotics. Sailors used to put silver coins in water barrels to keep water from contamination. Today's astronauts use silver in their water systems to sterilize recycled water.

Metalsmith Kathrin Schoenke notes too that because of the bacteria killing properties of silver, babies (usually of wealthy parents) used to be fed with the "silver spoon" to ensure that only germs-free food was given to the child. Before stainless steel could be formed and shaped by industrial methods, surgical instruments were also made of silver for the same bacteria killing reasons

For further information on the history and uses of silver, consult:

The Silver Institute, www.silverinstitute.org

The Silver Users Association, www.silverusersassociation.org

Fig. 5. 9 Sterling silver and stainless steel cheese knives by Louise Walker with a readily visible British Hallmark on the silver portion of the knife. In the U.K., any item weighing 7.78 grams or more must be hallmarked by an assay office in order to be sold as silver. *Photo by Louise Walker.*

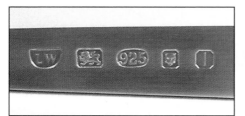

Fig. 5.10 Hallmark on silver knife above—makers mark of "LW" for Louise Walker, left facing lion passant for sterling silver, 925 fineness mark, London Assay Office mark (leopard's head without crown), date letter for 2010. *Photo by Louise Walker.*

Fig. 5.11 The stainless steel portion of the knife in figure 5.9 is laser marked "metal" to distinguish it from the hallmarked silver part. *Photo by Louise Walker.*

Silver Fineness Standards

US laws on the marking of precious metals started with the "National Gold & Silver Stamping Act of 1906." Years later, the Federal Trade Commission offered further guidelines stating with regard to silver, that "It is unfair or deceptive to mark, describe, or otherwise represent all or part of an industry product as "silver," "solid silver," "STERLING SILVER," "STERLING," or the abbreviation "STER" unless it is at least 925/1,000ths pure silver." For details and allowed markings and tolerances go to:

http://www.ftc.gov/bcp/guides/jewel-gd.shtm (Federal Trade Commission Guidelines)

http://uscode.house.gov/download/pls/15C8.txt (US law based on the National Gold & Silver Stamping Act of 1906.)

Sterling 925 is now the world's most common silver standard. However, it is not a universal standard for silver. According to the World Jewellery Confederation (CIBJO), an object must have 800 parts per thousand of silver by weight to be identified as a silver piece. However, in order to be identified as sterling silver, a piece must have at least 92.5% silver content. Table 5.1 presents other silver standards that exist now or have been used in the past.

The laws and standards of countries can take precedence over the CIBJO guidelines. For example, Belgium and Bulgaria recognize and have hallmarks for silver items that are for only 500/1000 silver. Romania's lowest standard for silver is 750.

Table 5.1 Silver content and notation*

Name	Silver %	Fineness Stamp	Notes
Fine silver	99.9%	999 or 1000	Pure silver
Britannia silver	95.84%	958	Formerly the British silver standard
French First Standard	95.0%	950	Popular in Peru and Thailand; French silver standard before 1973
Sterling silver	92.5%	925	Most common silver alloy fineness throughout the world; the silver standard now for the UK, US, Canada, and many other countries.
88 zolotnik Russian silver	91.66%	916	Used in the former Soviet Union from the 17th century until 1927
Coin silver	90.0%	900	A former standard for silver coins
84 zolotnik Russian silver	87.5%	875	Used in the former Soviet Union from the 17th century until 1927.
	83.5%	835	Used in Australia, Austria, Czech Republic, Belgium, Germany, Hungary, Israel, Portugal, Slovakia
"Scandinavian silver"	83%	830	Used in Bahrain, Lithuania, Finland, Norway, Portugal, Romania
"European Silver," "Egyptian silver"	80%	800	Used in Australia, Austria, Estonia, Hungary, Finland, Germany, Hungary, Ireland, Israel, Italy, Latvia, Lithuania, Malta, Poland, Portugal, Spain, Saudi Arabia, Slovakia, Sweden, Switzerland, The Netherlands, UK

* The countries listed in the notes are simply examples, not a complete list.
Sources for table and preceding text include: www.ftc.gov/bcp/guides/jewel-gd.shtm; www.cibjo.org; *CIBJO Precious Metals Book*; *World Hallmarks, Volume 1, 2nd Edition* by William B. Whetstone, Danusia V. Niklewicz, & Lindy L Matula; GIA *Gold & Precious Metals Course, Assignment 7.*

The varying hallmarking laws within Europe have created barriers to the international trade of precious metals. As a result, in 1972, seven European countries decided to simplify the requirements for trade between their respective countries. *World Hallmarks: Volume 1* (pp 347-351), published by the Hallmark Research Institute, states that the seven countries of Austria, Finland, Norway, Portugal, Sweden, Switzerland and the United Kingdom agreed to sign a treaty in 1975 and become members of what is properly known as the Convention on the Control and Marking of Articles of Precious Metals, also known as the **Hallmarking Convention** or the Vienna Convention. The treaty proposed the concept of using a Common Control Mark by all the Convention Members and to adhere, for example, to the strict minimum standard of 800 for silver and 830, 925, and 999 as the other standards for silver.

Sterling silver
Common Control
Mark

The success of the Hallmarking Convention is growing. To date there are a total of 21 ratified members. In 2005, Israel was the first non-European country to become a member. Other countries such as Canada, China and India closely follow the developments of the Hallmarking convention.

Metal and fineness marks for silver in the U.K. The metal type is indicated by the shape of the surround.

Avoiding Silver Tarnish

A disadvantage of most silver alloys is that they tarnish when exposed to sulfur elements in the air. **Tarnishing** is a surface discoloration and mainly results from the formation of silver sulfide, Ag_2S, although other compounds such as sulfate and chloride can also contribute to tarnishing.

Germanium is a silver-white element chemically similar to tin. In the 1990's, Peter Johns, a silversmith in England, discovered that the addition of germanium to silver could make it resistant to tarnish and fire scale. (**Fire scale** — also called "fire stain" — is a dark coating that forms on silver when the metal oxidizes at high temperatures.) This tarnish-resistant silver is patented and sold under the trade name **Argentium**.™ There are two grades of Argentium silver: 93.5% minimum silver content and 96% minimum silver, which raises the level above the UK Britannia standard. Argentium International Ltd., guarantees that all of its silver is recycled and nickel free. United Precious Metal Refining (UPMR) in Alden, NY has also developed and patented a germanium-based tarnish resistant silver, which they call **Sterlium Plus**. Generic tarnish resistant silver containing germanium is also available from casting houses and refiners; jewelers find it very helpful for preventing tarnish problems.

Other tarnish resistant alloys have been created by replacing the 7.5% copper in sterling silver with materials such as platinum, palladium, silicon, zinc or nickel. The presence of copper in sterling silver accelerates tarnishing, so when copper is omitted, the resulting silver alloy is more tarnish resistant. However, it often tends to be softer than and not as strong as sterling silver made with copper. **Platinum sterling** is marketed as a lower cost alternative to white gold by ABI Precious Metals. United Precious Metals has a platinum sterling which they have trademarked "Platinet." **Palladium sterling**, produced with varying amounts of palladium, is available from a wider number of producers because of the lower cost of the palladium.

Tarnish-resistant silver is often sold as "tarnish-free" silver, but this can be misleading. Any silver product can tarnish over the long term.

Fig. 5.14 Anti-tarnish half clear zip lock bag. *Pouch and photo from Intercept Silver and Jewelry Care Company.*

Fig. 5.15 Anti-tarnish tabs for containers with silver items. *Tabs and photo from Intercept Silver and Jewelry Care Company.*

Another way to make silver tarnish-resistant is to plate it with rhodium, platinum or palladium or to coat it with a non-metallic substance. Plated or coated silver may cost less than tarnish resistant silver, but the plating or coating can wear away or come off when the silver is polished and/or worked on by a jeweler. This can happen, for example, when any ring sizing or soldering is performed. The resulting exposed metal can tarnish. If your goal is to buy silver that is inherently tarnish resistant, ask first if the silver is plated or coated. For more information on platings and coatings, see Chapter 13.

Chemical solutions can also help protect silver from tarnishing. Legor makes an anti-tarnish solution that can be used when a regular sterling silver piece has been completed. It is dipped into the bath and the ingredients in the chemical fill the fine pores of the silver surface and seal it for some time against tarnishing.

If you have jewelry or objects that are not tarnish resistant, plated or coated, they can be protected by storing them with sulfur absorption strips or in pouches, bags or containers designed to absorb sulfur (figs, 5.8 & 5.9) for a specified number of years. Lucent Bell Labs' restoration of corroded copper areas on the Statue of Liberty in the mid 1980's helped Lucent discover nontoxic sulfur absorption materials. Curiously, figuring out a way to provide green replacement copper to match the green patina of the weathered copper on the rest of the statue led Bell Labs to find a method of preventing the formation of tarnish on silver, which is now a patented process called Intercept Technology™. Volkswagen uses Intercept Technology™ packaging to prevent corrosion of its car parts and engines during shipping and storage; many museums and jewelers use Intercept Technology™ strips and filtration units for display cases and bags for storage to prevent metals from tarnishing.

Removing Silver Tarnish

If your silver is not tarnish resistant, plated, coated or protected with sulfur-absorbing material, it may have darkened after being exposed to corrosive agents in the air. The silver can be returned to its former luster by removing the silver sulfide. There are two ways to remove it. One way is to remove the silver sulfide from the surface of the silver mechanically (e.g., buffing), but some silver will also be removed.

The other way to remove tarnish is to reverse the chemical reaction and turn silver sulfide back into silver by dipping the silver object into a chemical solution. In this process, the silver remains in place.

Here's a simple way for you to conduct this electrolytic-type process, which will allow you to only remove the sulfate but leave the silver intact. You'll need a sheet of aluminum foil, hot water, baking soda, salt and a pot or pan. Then:

1) Line the pot or pan with the aluminum foil and place your tarnished silver jewelry item(s) or object(s) on top of the foil. **The object must have direct contact with the aluminum foil** for the process to work.

2) Pour hot water over the item so that it is entirely covered. Add a tablespoon of salt (two for a large pot) to create the electrolytic solution.

3) Pour a package of fresh (FRESH!) baking soda into the mix

4) Let it sit and watch how the sulfate is stripped off the silver surface. Remove the items when they are clean.

Naturally this process only works for silver items without pearls or other gems that should not be exposed to hot chemical solutions. Remember also that hot water can adversely affect the life span of glue.

This process works because aluminum has a greater affinity for sulfur than silver does. In the above process, the silver sulfide reacts with aluminum and sulfur atoms are transferred from silver to aluminum, freeing the silver metal and forming aluminum sulfide. Chemists represent this reaction with a chemical equation:

$$\text{silver sulfide} + \text{aluminum} = \text{silver} + \text{aluminum sulfide}$$

Silver alloys can differ in composition and may have tarnish layers other than silver sulfide. Consequently, the above method might not be adequate for removing all tarnishes from silver. Chemical shelf cleaners may be required, especially for silver that is below 925 in purity.

The above information on removing tarnish is from Kathrin Schoenke, a jeweler and platinum industry consultant.

Debra Sawatzky, an appraiser specializing in antique and estate jewelry, advises that new cleaning techniques be reserved for new items and that old items be cleaned by hand, the old-fashioned way. Otherwise you may devalue the older items, including flatware and hollowware. She says that people pay premium prices for patina, the surface appearance of silver grown beautiful with age, and that the glow from hand polishing is more desirable on old objects than that obtained with modern cleaning methods.

Oxidized Silver

The previous sections discussed ways to avoid oxidation and the resulting tarnish. This section briefly discusses ways jewelry is intentionally oxidized to create a dark color (patina) on metal. Two technical terms for darkening or coloring metal are **oxidizing** and **patinating**. Jewelry with oxidized silver or other metals is becoming more and more popular because the contrast between dark and bright areas on jewelry allows for unique designs. In addition, it eliminates tarnish concerns.

Fig. 5.16 Oxidized sterling silver bracelets by So Young Park. *Photo by So Young Park.*

Fig. 5.17 Oxidized sterling silver & 18K gold pendant by Todd Reed. *Photo by Brian Mark.*

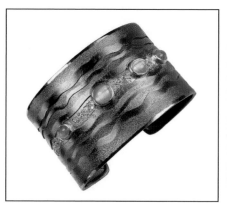

Fig. 5.18 Oxidized sterling silver, 24K, 22K & cat's-eye opal bracelet design by Peter Schmid. *Jewelry: Atelier Zobel; photo by Fred Thomas.*

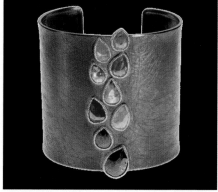

Fig. 5.19 Oxidized sterling silver, 18K gold and diamond bracelet by Todd Reed. *Photo by Brian Mark.*

Fig. 5.20 Sterling silver name pins from the collection of the Birmingham Assay Office in Britain. Dating from 1880-1920, these pins were the first type of jewelry available to the working class in the U.K. The office has collected silver items dating back to the 17th century because they can be used to help determine the age of antique silver objects with XRF analysis. The composition of the silver alloys changed depending on the period in which they were manufactured.

Photo by Craig O'Donnell.

The disadvantage of patinized jewelry is that scratches and scrapes are more noticeable on oxidized surfaces, which is one reason that oxidation is sometimes limited to recessed areas of a jewelry piece. However, entire surfaces of silver mountings are also being oxidized. These can be coated with colorless substances to help prevent removal of the surface coloring.

Here are some ways that silver and some other metals can be oxidized.

Liver of sulfur method: Immersion in a liver of sulfur (potassium sulfide) solution is the most common way to oxidize silver. In his book, *Contemporary Jewelry* (p 280), Phillip Morton says that the solution can be made by dropping one or two lumps of potassium sulphate into a pint of hot water. The solution must be kept in a tightly closed jar or bottle, since oxygen will cause it to deteriorate. Place the article into the jar for a period of time, from a minute to a half an hour depending on the darkness of color desired. The longer the metal is left in the solution, the thicker the coating of oxide. Metal that is to be oxidized should be completely cleaned beforehand.

In *Jewelry: Fundamentals of Metalsmithing* (pp 65-66), Tim McCreight suggests "dip the piece quickly [in the liver of sulfur solution], then rinse it in running water to check the effect. Before turning gray, sterling progresses though these colors: yellow, brown, crimson, and electric blue. Though none are as durable as the gray, these colors can last quite awhile on pins and earrings."

Jinks McGrath says in the *Encyclopedia of Jewelry Techniques* (p 94), "if liver of sulfur is used in a hot solution, standard silver and golds that are less than 18 carat will turn black. A few drops of household ammonia can be added to give an even deeper black . . . When it is used cold, liver of sulfur will color copper, but it will have little or an unpleasant effect on silver."

Chloride of platinum method: Silver can also be oxidized by applying a solution of chloride of platinum (1 ounce chloride of platinum to 1 gallon of water) onto a silver piece with a brush. Only a small quantity is required. (*Contemporary Jewelry*, p. 280)

Torch method: An oxidizing flame is used to heat the metal. The heat from the flame combines with the oxygen in the air to form an oxide film on the surface of the metal. However, it is usually difficult to control this process.

Silver jewelry is reasonably priced, versatile and the most reflective of all the precious metals. Silver can go with any outfit; it can be casual, yet worn with dressy clothes. However, silver is facing competition from other affordable white metals. They are discussed in Chapters 6 and 10.

6

Copper, Brass & Bronze

Copper was the first metal used in jewelry because it was available in large quantities and was found almost at the surface of the ground. Archeological excavations have proved that copper crafting was known in Iraq and Iran more than 10,000 years ago.

In the Roman era (27BC to 476AD), copper was primarily mined on Cyprus, so it was called *cyprium* (metal of Cyprus), later shortened to *cuprum* and abbreviated as "**Cu**" for its chemical symbol. Pure copper is reddish brown, and when combined with other metal(s), it can turn the alloy pink, reddish, orange or yellow. The only other elemental metals that are not gray to white are gold, which is yellow and osmium, which is bluish. Currently, the largest copper deposits are in Russia, Chile, Namibia, the Congo and the United States.

Copper is an ideal metal for making ornaments and tools because it's attractive, malleable, ductile, and resilient. One drawback is that it not only discolors when worn but also discolors the wearer as well—usually by turning the skin green. The green coating that forms on copper and its alloys on long exposure to moisture is called a **patina**. Despite its tendency to turn green, copper is generally considered corrosion resistant. In fact, copper-containing paints are used for the hulls of ships because the copper resists corrosion from seawater and doesn't rust. In addition, the copper reduces the growth of marine weed, algae and barnacles on the hull, thereby allowing the ship to move faster through the water.

Besides being important for the electronics industry because of its high thermal and electrical conductivity, copper is becoming increasingly important for the prevention of infection because of its antimicrobial properties. The US Environmental Protection Agency has approved the registration of more than 350 antimicrobial copper alloys because laboratory tests as of 2011 have shown that when cleaned regularly, copper, brass and bronze kill more than 99.9% of staph and e, *coli* bacteria within two hours of exposure. The amount of disease-producing bacteria in healthcare facilities could be reduced if these copper alloys were used more for door and furniture hardware, bed rails, intravenous (IV) stands, faucets, sinks, etc. Unlike antimicrobial coatings, the antibacterial efficacy of copper metals won't wear away.

Copper is an essential element in the diet found especially in foods such as nuts, liver, seafood, chocolate and legumes; it's important for growth, the immune system, tissue repair, the heart, brain, bones, and the formation of red blood cells. As a result, many crystal healers believe that wearing copper jewelry offers health benefits. However, if pure copper is ingested, it can be toxic. In addition, if copper is heated and forms cupric oxide, the resulting vapor can irritate the lungs, skin and eyes. Even though copper cookware is used by many professional chefs because of its exceptional heat conductivity, it is lined with other metal such as stainless steel to prevent the copper from reacting to the food, causing nausea, vomiting and diarrhea.

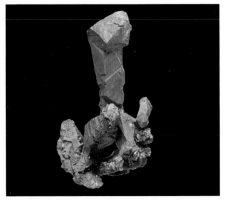

Fig. 6.1 Copper from Siberia Russia. *Specimen from Pala International; photo: John McLean.*

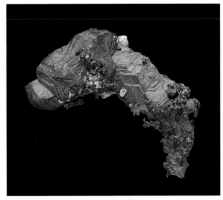

Fig. 6.2 Copper from Michigan. *Specimen from Pala International; photo: John McLean.*

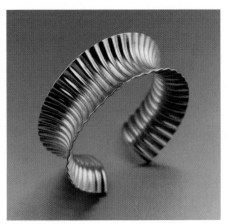

Fig. 6.3 Copper bracelet handcrafted by John S. Brana. *Photo by John S. Brana.*

Fig. 6.4 Brass (Nugold) bracelet handcrafted by John S. Brana. *Photo by John S. Brana.*

Since the price of gold has risen, more copper and copper alloys are being used and created for jewelry mountings, clasps and other findings, both because of their much lower cost and their beauty. When copper is mixed with other elemental metals such as zinc, tin, silicon and nickel, it becomes harder and stronger and therefore more resistant to wear. The next sections briefly discuss copper alloys such as brass, bronze and shibuichi and show how they are being used in jewelry.

Brass

Brass is an alloy of copper and zinc, but the proportions can vary; other elements such as silicon, aluminum and manganese may also be present to make it more corrosion resistant. According to Steve McCreight, in *Fundamentals of Metalsmithing,* the most common mix is called yellow brass or CDA 260, and consists of 30% zinc and 70% copper. As the proportion of copper increases, the color becomes more golden and the metal more malleable.

NuGold is another common brass alloy, which is often used in jewelry making and is nickel-free. The name is also written as "Nugold," "nugold," or "Nu-gold." It contains about 85–88% copper and the balance is zinc. The Copper Development Association (CDA) identified it as #230. The three-digit CDA identification numbers such as the ones for NuGold and yellow brass have been expanded to five digits following the prefix letter C for copper and made part of the Unified Numbering System for Metals and Alloys. UNS designations are simply expansions that supercede the former CDA designations. For example, the UNS number for NuGold is C23000. This numbering system is also used in Canada.

Brass is often used in ethnic and hand-crafted jewelry. Unfortunately it turns green when worn next to the skin, so it's not a good choice for rings or chains unless it is plated or coated. Brass ear wires or posts should not be worn in pierced ears.

Bronze

Bronze is the oldest alloy known to man, dating back to about 3500 B.C. when it was used to make arrowheads and spears. Traditionally bronze has been defined as a copper alloy containing tin. Today, however, bronzes are considered to be copper alloys in which the major alloying element is not zinc or nickel (source of new definition: www.copper.org).

Table 6.1 Brass Versus Bronze

Brass	Bronze
Copper alloy that contains zinc as the principal alloying element with or without other alloying elements such as silicon, iron, aluminum and nickel.	Originally "bronze" described copper alloys with tin as the only or principal alloying element. Today, bronzes are copper alloys in which the major alloying element is not zinc or nickel. Source of definition: www.copper.org
Normally has a golden or subdued yellow color	May be reddish or greenish brown, golden, or grayish depending on the alloy.
Often used as a base metal for gold-plated jewelry and enamels.	Recently used as a gold substitute in jewelry; ideal for casting
Usually has a lower melting point than bronze—around 950°C	Usually has a higher melting point than brass—a little above 1000°C
Typically softer than bronze	Typically harder than brass
Often used for musical instruments, locks, plumbing, doorknobs, valves, electrical applications, clocks and decorative items	Commonly used for bells, statues, springs, ship propellers, weatherstripping

Sources include: *GIA Jeweler's Manual*; *Jewelry: Fundamentals of Metalsmithing* by Tim McCreight, *Jewellery Technology* by Diego Pinton, www.copper.org, www.diffen.com, www.patel.mech.com, www.wikipedia.org.

Fig. 6.5 Cold-forged brass (nugold 85-15) and anodized titanium earrings by Fred & Kate Pearce. *Photo courtesy Pearce Design.*

Fig. 6.6 Bronze magnetic clasp and photo by Kim Fox of Handfast Design.

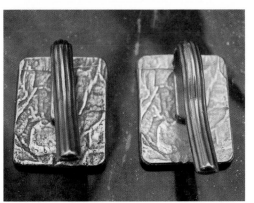

Fig. 6.7 White bronze and jewelers bronze bail clip and photo by Kim Fox, owner of Handfast Design.

Fig. 6.8 Brass and sterling silver necklace by Fred & Kate Pearce. *Photo: Pearce Design.*

Most bronze is a golden to reddish brown color but it can also resemble silver if it is alloyed with a high percentage (e.g., 40%) of white metals such as magnesium, nickel, silicon or zinc. Figure 6.7 shows an example of a white bronze bail clip next to one of jewelers bronze. A new trend in jewelry is to use bronze as a gold substitute in fine jewelry.

Table 6.1 compares bronze to brass, and Table 6.2 provides information about the composition, melting points and specific gravity of various brass and bronze alloys.

Fig. 6.9 Bronze and amethyst ring and photo by John S. Brana.

Table 6.2 Alloys Containing Copper and other Elements*

Name	Specific Gravity	% Ag	% Cu	% Zn	% Other	Melting Point	
						°C	°F
Brass, cartridge #260	8.5		70	30		954	1749
Red brass	8.8		90	10	some contains lead and/or tin	1044	1910
Yellow brass #260	8.5		66	33		955	1750
Gilding metal brass			95	5			
NuGold #230			85–88	15-13			
Bronze	8.8		96		4 tin	1060	1945
Jewelers bronze # 226	8.7		88	12		1030	1886
Silicon bronze			97–98		2–3 silicon		
White bronze			≈60		≈ 40 white metals such zn, mg, ni,		
Copper (Cu)	8.9		100			1083	1981
Nickel (Ni)	8.8				100 nickel	1455	2651
German silver	8.8		65	23	12 nickel	1037	1900
Nickel silver #752	8.8		65	17	18 nickel	1110	2030
Silver (Ag) (fine)	10.6	100				961	1762
Sterling silver (925)	10.4	92.5	7.5			920	1640
Coin silver (800) or European Silver	10.3	80	20			890	1634
Shakudo			96		4 gold		
Shibuichi (pink silver)		usually 25 or less	usually 75 or more				
Tin (Sn)	7.3				100 tin	232	450
Zinc (Zn)	7.1			100		419	786

* Please note that these are standard idealized alloys; bronze, for example, can contain a variety of other metals besides tin.

Sources of data: *Jewelry: Fundamentals of Metalsmithing* by Tim McCreight, *Professional Goldsmithing* by Alan Revere.

Fig. 6.10 From left: shibuichi (pink silver), rear: sterling silver, right: silicon bronze, front: white bronze. *Magnetic clasps and photo by Kim Fox owner of Handfast Design.*

Fig. 6.11 Bronze and amber ring by John S. Brana. *Photo by John Brana.*

Fig. 6.12 Bronze and pure 24K gold necklace from Gurhan's Museum collection. *Photo by Gurhan.*

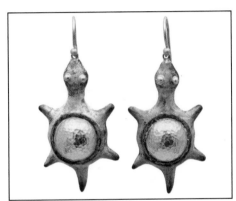

Fig. 6.13 Bronze and pure 24K gold turtle earrings from Gurhan's Pantheon Collection. *Photo by Gurhan.*

Fig. 6.14 Nugold brass cuff bracelet by John S. Brana. *Photo by John S. Brana.*

Fig. 6.15 Copper earrings by John S. Brana. *Photo by John S. Brana.*

Shibuichi

Shibuichi is a traditional copper/silver alloy that is pinkish. It means "one-fourth" in Japanese in reference to its historical composition of 1/4 silver and 3/4 copper. Other ratios are also used today such as 5% silver and 95% copper. Long ago, shibuichi was primarily used to make ornaments for samurai swords. Now it is commonly used in mokume gane metal (a layered metal resembling wood) or to create metal with subtle green and blue patinas. Because of the rise in the price of silver and gold, shibuichi is also being crafted by itself as an alternative jewelry metal. Figure 6.10 shows an example of a shibuichi clasp next to three others made of sterling silver, white bronze and jewelers bronze.

Identifying Copper, Brass and Bronze

Copper is the only reddish brown metal, so it is easy to identify uncoated pure copper by color. In addition, whenever a non-plated and non-coated alloy is pink or reddish, copper is a component. Brass generally has a grayish yellow or golden color, but bronze can vary in color depending on its composition and the degree to which it has been exposed to oxidation. More often than not it is reddish brown to golden in color; it may also resemble silver if it contains a high percentage of white metals. Like copper, bronze may be greenish if it has been exposed to a lot of weathering.

Copper is not magnetic and neither are alloying metals such as zinc, tin, silver, magnesium, silicon, aluminum, gold or lead. Therefore, a magnet is an ideal tool for separating brass, copper, bronze and precious metals from plated or unplated steel, or iron. Steel is often plated with gold and may resemble gold or brass. Copper is more expensive than steel; during 2010 and 2011, copper ranged from about $3.00 to $4.40/pound whereas steel typically sells for less than 50 cents per pound and is usually sold by the ton for much less. The exact price depends on the type, shape and amount of steel being purchased.

If you have a pile of unidentified jewelry or metal, you can save time by first checking it with a magnet and separating the magnetic pieces from those that are not attracted to the magnet. Large crane-held magnets are used in scrap yards and metal recycling facilities to sort the steel.

Check for markings on the piece with a loupe or other magnifier. Copper, brass and bronze are seldom identified on the piece, but gold, silver, platinum and palladium are. If the marking indicates the piece is composed of precious metal, further testing is necessary. See Chapter 8, "Real or Fake?" for information on touchstone and acid testing and gold and platinum testing machines. The chapter includes photos of two x-ray fluorescence devices that can indicate all the alloying metals and their percentages in a jewelry piece. Chapter 9 will help you determine the fineness of gold.

7

Manufacturing Methods

Imagine being able to create jewelry from a computer-aided design (**CAD**) with a machine similar to an inkjet type printer. This technology is already available and is called **3D** (or **3d**) **printing** or **additive manufacturing** because it involves adding layer upon layer of material to create a three-dimensional object. This process is also identified as direct digital manufacturing (**DDM**).

CAD can also be used for machining solid objects with a milling machine that automatically cuts away an item from wax, metal or other substances. This is referred to as **milling** or **subtractive manufacturing.** The machine that creates the object is called a mill, milling machine or computer numerical control machine (**CNC machine**).

Besides discussing 3D printing and milling, this chapter presents four traditional methods of making jewelry: lost wax casting, stamping (die-striking), electroforming and hand fabrication. In addition, it explains the benefits of each method and how combining the methods may be advantageous when producing a piece.

Lost Wax Casting

Lost wax casting dates back to at least 1500 BC, when the Egyptians used it. For a while this method disappeared, but in recent years it has become the most widely used manufacturing process. Lost wax casting involves a series of steps, as follows:

1. Usually a wax model is created by hand or with a milling machine. A proprietary resin model can be created with 3D printing. Computer-aided-designs (CAD) can be used to automatically create models with either the mill or the 3D printer. Sometimes the model is a piece that has already been manufactured in metal.

2. The wax model is attached to wax rods called **sprue rods**; each one forms a channel to allow the wax to be melted out and molten metal poured in.

3. In the case of wax and resin models, a metal canister called a flask is then placed over the wax and into the base. It is important that no bubbles adhere to the model, so it is often painted with a de-bubblizing solution before a liquid plaster called **investment** is poured into the canister.

4. Air is vacuumed out of the newly mixed investment.

5. The investment is then poured into the flask and allowed to harden.

6. The high-temperature liquid plaster is heated, and the wax or proprietary resin melts and pours out of a hole. A hollow plaster mold is left. The lost wax process gets its name from the fact that the model is "lost" through melting as the flask is heated.

The flask must be heated long enough to insure all traces of the model are burned away. It then must be maintained at a high temperature to insure that the molten metal does not "chill" when cast into the mold.

7. Molten metal balls up (like mercury) so either vacuum or centrifugal force is used to force it into the mold. "Sling" casting (swinging the mold to create centrifugal force) is used by artisans in parts of the world without access to electricity.

8 The mold is broken and the master model is removed and cleaned up.

9. At this point a rubber mold of the metal model is made (fig. 7.1). The model is placed in a rectangular frame and rubber is packed around it, the rubber is heated so that it flows around the model and is vulcanized into a solid block. The mold is cut and the master removed leaving a hollow mold. In addition to vulcanized rubber molds, molds are now often made with liquid silicone products that don't have to be heated. This allows for "safety molds" of wax originals to be made before casting.

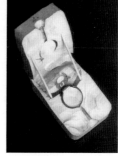

10. Wax copies can be made by injecting melted wax into the mold through a hole.

Fig. 7.1 Rubber mold

11. Wax copies can be linked like branches of a tree so that multiples can be cast at one time (fig. 7.2).

12. The tree of waxes is covered with a high-temperature plaster (investment). Once again heat is applied and a hollow mold is created. Molten metal is poured into the hollow plaster mold the hollow plaster mold is then broken and the pieces are removed, separated from the tree and cleaned up and polished.

The steps of original wax or 3d model to master, master to mold and mold to wax, then wax to finished piece have a cumulative shrinkage of about 5 to 9%. The original wax is bigger than the finished piece. Designers take that into account when they create.

An advantage of using models created from CAD designs by milling machines or 3D printers instead of waxes created from a mold is that every piece made from the CAD model is much closer to the original size.

Advantages of casting

♦ It's a relatively quick way of making several identical pieces.

♦ It offers unlimited design possibilities. You can even draw your own designs and have the jeweler transfer them to a wax model.

♦ It's economical when many pieces are produced from the same mold. One-of-a-kind cast pieces, however, can be just as costly as those which are hand-fabricated.

♦ It's an easy way to make copies or matching pieces. For example, a necklace might be used to make a mold for a matching bracelet. Hanna Cook-Wallace of Studio Jewelers Ltd. points out that CAD allows you to have an heirloom piece reproduced if it is beyond repair (or copies are wanted by family members). Coordinates based on the original piece are taken, and the piece can be rebuilt in the rendering before the 3D mold is produced.

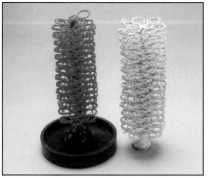

Fig. 7.2 "Tree" with wax models and the resulting gold castings. *Photo from Stamper Black Hills Gold Jewelry.*

Fig.7.3 Cast Deco Bloom sterling silver clasp by Kim Fox, owner HandFast Design; *photo by Robert Liu.*

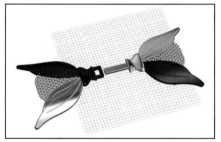

Fig. 7.4 CAD rendering of Deco Bloom clasp in fig. 7.3. *Clasp & CAD design by Kim Fox.*

Fig. 7.5 Wax for Deco Bloom clasp in figure 7.3. *Wax & CAD design by Kim Fox.*

Figs. 7.6–7.8 CAD rendering, wax, and finished magnetic sunflower bronze clasp. *Clasp, CAD designs and photo by Kim Fox, owner of HandFast Design.*

Figs. 7.9–7.11 CAD rendering, wax, and finished ring. It is usually easier for customers to obtain design options and visualize the final jewelry piece with a CAD image than with a sketch or wax. *Gold ring and photo by Kim Fox, owner of HandFast Design; ring courtesy Elaine Ferrari.*

♦ Your original drawings may be used by your jeweler, who can help you refine your design to ensure a product that is durable—and wearable. CAD renderings will show you several views of a piece before the wax is built. Expect to pay a non-refundable design fee for custom design work.

Disadvantages of casting

♦ Cast metal is usually less suitable for fine engraving because it tends to be more porous than stamped and hand-fabricated pieces.

♦ Casting flaws can include porosity and loss of detail (however, today's laser welding technology allows for the complete repair of porosity).

♦ Casting usually requires more cleanup and finishing than stamping, electroforming and fabrication. As a result, cast jewelry is often rougher and duller underneath, especially in hard to reach areas. Many CAD designs don't take clean up (or other practical concerns) into consideration, so these can be quite problematic.

Stamping (Die-Striking)

Another ancient technique for making jewelry is **stamping**, which in the United States is also called **die-striking**. In this process, metal is punched between two carved metal blocks called **dies**, creating a form and design. The metal becomes very dense and strong as the hydraulic presses squeeze it between the dies at pressures of up to 25 tons per square inch.

The quality of stamped jewelry has greatly improved over the years. In some cases, it even looks like it's hand fabricated. The stamping process is often used to make earrings, pendants, coins, wedding bands (fig. 7.12), settings, and fancy Italian chains.

Fig. 7.12 Die-struck ring and photo from True Knots.

Advantages of stamping

♦ Once the dies are made, it's a faster process than other methods.

♦ The quality is consistent from one piece to another. Edges are perfectly straight, and shapes are perfectly symmetrical. Consistency may also be achieved by using machines to cut and finish cast pieces.

♦ In large quantities, stamping is economical.

♦ Stamped items usually wear better than cast items, because of their high density.

♦ Stamped pieces require little cleanup before polishing.

♦ Stamped jewelry can take a very high polish due to its density.

♦ Stamped pieces can be thinner than those which are cast. Therefore, they may provide a bigger look at a lower price.

♦ Stamped metal is ideal for engraving because of its high density.

Disadvantages of stamping

♦ It's not suitable for small-scale manufacturing or limited runs, because of the high cost of the dies, equipment and set up.

♦ It's a longer process than casting if the dies are not readily available. To have dies made may take a month and is expensive.

♦ Design options are limited.

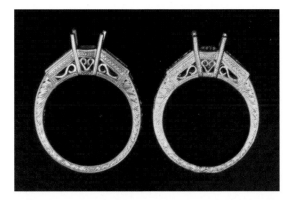

Fig. 7.13 Hand-fabricated, hand-engraved ring by Varna Platinum (left), which was used to make a mold for the cast platinum ring on the right. The hard to reach areas in the cast ring are rougher, duller and shallower. The metal in the hand-fabricated ring is denser, in contrast to the more porous cast ring; and the engraving is crisper and smoother. Normally, casting is not a good manufacturing option for fine engraved pieces.

Fig. 7.14 The difference between the hand-fabricated ring (left) and the cast ring on the right becomes more obvious when viewed close up.

All photos this page copyright by Renée Newman.

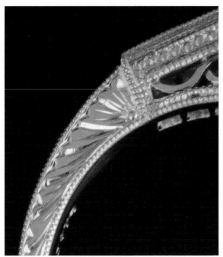

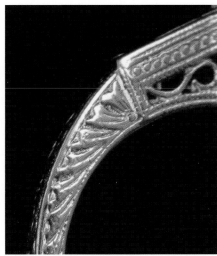

Fig. 7.15 Hand-fabricated ring by Varna viewed under a 10-power microscope

Fig. 7.16 Cast ring under 10x magnification

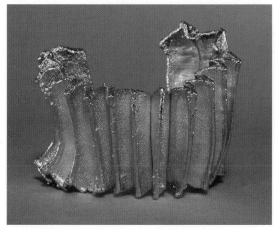

Fig. 7.17 Electroformed sterling silver earrings and image from Metal Marketplace International.

Fig. 7.18 A combination fabricated (foldformed) and electroformed bracelet, first in copper and electroformed in 24k gold. *Bracelet and photo by Charles Lewton-Brain.*

Electroforming

Electroforming is a technique of forming metal objects by electrically depositing the metal over a mold called a **mandrel**. The mandrel, which may be made of a material such as wax, epoxy resin, silicone rubber, stainless steel or even a baby shoe, is usually later removed, leaving a metal shell. However, the mandrel may be left in the piece, becoming part of it as in the pleated (fold-formed) bracelet by Charles Lewton-Brain shown in figure 7.18. It was electroformed in 24K gold over hand-fabricated copper.

The process of electroforming originated in the 1830's with the birth of electroplating (Rod Edwards, *The Technique of Jewelry*, p 202). It became a convenient and extremely accurate way for museums to reproduce antique pieces. Originally, gold electroforming required a very high karat gold—at least 23.5K. But in the early 1980's, a French jewelry manufacturer developed a way of making electroformed pieces out of 14K and 18K gold. Since then, electroformed jewelry has become increasingly popular. Today, more and more silver electroformed pieces are appearing on the market because of the high price of gold.

Sometimes electroforming is confused with electroplating. The biggest difference is in the thickness of the metal. Electroplating finishes are measured in millionths of an inch or in microns (thousandths of a millimeter. Legally, electroplating with 24K gold must be no less than seven-millionths of an inch. A layer that thin could never form a piece of jewelry. Another difference is that in electroplating, the cathode is usually a finished object getting a surface coating of precious metal. In electroforming, the mandrel is the cathode (from Assignment 9 of GIA's *Gold & Precious Metals*).

Advantages of electroformed jewelry

♦ It provides a big look at a relatively low price.

♦ It's lightweight and therefore ideal for earrings.

♦ It can show minute detail like one would see in a fine engraving.

Disadvantages of electroformed jewelry

♦ It dents very easily. The average thickness of the higher quality pieces is about .007", so special care is required. Electroforming would be an unsuitable process for making rings.

♦ It cannot be repaired. Some manufacturers, however, offer a lifetime replacement warranty.

♦ It usually costs more per gram than stamped and cast jewelry. This is because expensive equipment is used to make it, and it tends to be produced in lower quantities.

♦ Even though gems have been set in electroformed jewelry, the thinness of the metal often makes setting impossible.

♦ It's usually difficult to add a textured finish to electroformed jewelry because of the thinness of the metal.

Electroformed jewelry can vary in quality depending on the manufacturer. Some pieces may have a metal thickness of only .003 of an inch. Others may lack the bright polish or stronger post assembly of a better quality piece. The thinner, more fragile pieces may be cheaper; but if they don't last as long, they shouldn't be considered a better buy.

3D Printing from CAD (Additive Manufacturing)

3D printing involves turning a computer-aided design (**CAD**) into a three-dimensional physical object using a machine generically called a 3D printer—a machine that grows substances such as wax or metal. Instead of casting an object or taking a block of material and cutting away until an object is produced, a 3D printer builds the object layer by layer from microscopic particles in order to allow for a smooth surface. Since 3D printers create objects layer by layer, they can create items with internal movable parts, instead of having to print separate parts and have a person spend time assembling them.

3D metal printing was initially used to make tools and machinery. Now it is employed to make many other products such as cars, dental crowns, medical implants, shoes, resin models for casting and more recently jewelry. Biotech firms hope that someday 3D technology will even be able to create body parts and organs.

Fig. 7.19 A Solidscape 3D printer and photo from Romanoff International Supply Corp.

Fig. 7.20 Solidscape 3D printer creating six resin ring models. *Photo © Renée Newman.*

One 3D printer is the Solidscape, which is used for making investment casting models for the jewelry, medical, dental and industrial markets (fig 7.19). Using CAD files, it creates models with a proprietary resin instead of in wax in a period of several hours. Figure 7.20 shows six ring models being produced on a Solidscape. Depending on the model, the Solidscape sells for roughly between $30,000–$40,000.

Another 3D printer named the Concept Laser enables manufacturers to bypass the lost wax casting process by making jewelry directly without a wax or resin model. Jewelry was first made on this machine in Germany using an atomized 18K gold alloy powder produced by the Legor Group in Italy. In 2012, the Concept Laser was introduced to the American market by Romanoff International Supply Corporation on behalf of Hofman Innovation Group, the German company that produces the machine. The Concept Laser costs $289,000 and is currently being purchased either by very large companies or jewelers that sell a high volume of very expensive high-end jewelry. As of 2012, only atomized 18K gold powder was available because 18K is the gold fineness usually sold on the European continent. A 14K and 10K powder will be available in the near future.

3D printers can usually print objects in a few hours, but the time varies depending on the type of machine and the number of objects being produced simultaneously.

The 3D-printing process is a little different for metal jewelry than for wax models used for casting. You can actually see wax models being printed on a 3D printer (fig. 7.20), whereas the Concept Laser makes metal jewelry within an open box of atomized metal particles, which is placed on a moveable platform. A laser beam comes down into an open box of powder and melts the powder at the spot that equals the CAD designation while the platform gradually moves down, enabling another microscopic layer to be melted onto the piece. Since the jewelry is created within a powder, it is not possible to view the growing process. The powder must be brushed away from the piece(s) before the jewelry can be seen. According to Bob Romanoff, who is in charge of distributing the machine in America, the Concept Laser can create jewelry that is 99.4% porosity free because it is grown in a vacuum atmosphere with microscopic particles.

CAD Milling (Subtractive Manufacturing)

Most lost-wax casting models today are designed with CAD and are made on milling machines (also called mills). One of the newest mills is the Revo540CX produced by Gemvision and priced just less than $25,000. It has two high-speed spindles that approach the model from two different directions in the same run. This enables one to mill virtually any wax model in a single operation without the hassle of changing fixtures or part position; consequently, the set up is easier, models are more precise, and jobs finish faster than on mills with only one spindle. The Revo540CX doesn't produce jewelry directly yet. However, the milling process can and has been used to directly cut very basic metal jewelry designed with CAD. For example, Stephen Vincent in Minneapolis, Minnesota uses a mill to create wedding bands and basic rings from CAD (figs. 7.29–7.32.). Milling machines are not able yet to produce metal jewelry with intricate and complicated designs.

Fig. 7.21. Metal powder being removed from around the rings that were grown in a Concept Laser 3D printer. *Photo courtesy Romanoff International Supply Corporation.*

Fig. 7.23 Mlab Concept Laser 3D printer. *Photo and machine from Romanoff International Supply Corporation.*

Fig. 7.22 Closer view of rings in 3D printer after metal powder removal. *Photo by Renée Newman.*

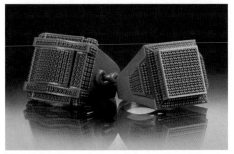

Fig. 7.24 Wax ring models produced by the Solidscape 3D printer. Note the fine detail and regular prongs. *Models and photograph courtesy of Solidscape.*

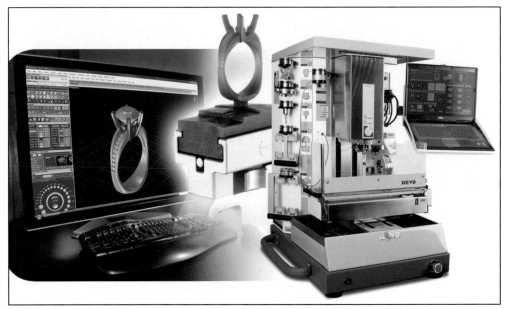

Fig 7.25 Gemvision Revo540C mill used to produce wax models from CAD designs. *Photo courtesy Gemvision.*

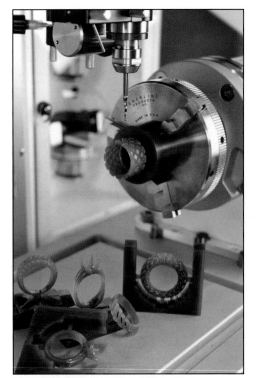

Fig. 7.26 Wax model created by a mill from a CAD design. *Photo courtesy Gemvision.*

Fig. 7.27 Milling machine drill cutting away wax to form a ring from a CAD design, with wax models below. *Photo courtesy Gemvision.*

Fig. 7.28 Wax model created by a mill from a CAD design. *Photo courtesy Gemvision.*

Fig 7.29 Milling machine used by master goldsmith Stephen Vincent to produce metal jewelry from his CAD designs. *Photo courtesy Stephen Vincent Design.*

Fig. 7.30 Stephen Vincent viewing his CAD designs on a monitor. *Photo by Louisa Marion Photography.*

Fig. 7.31 Time exposure showing milling drill cutting away metal to form a ring from a CAD design by Stephen Vincent. *Photo by Louisa Marion Photography.*

Fig. 7.32 Stephen Vincent Design ring produced by the milling machine in figure 7.29. *Photo coutesy Stephen Vincent Design.*

Advantages of using CAD and 3D printing or milling for models

♦ It's a relatively quick way of making a few or several identical pieces.

♦ It offers unlimited design possibilities. You can even draw your own designs and have them transferred to a computer model. Complicated designs are best done on a 3D printer.

♦ It enables you to better visualize the finished piece than with a sketch or wax model.

♦ It is easier to make changes on a CAD model than on a wax or metal model.

♦ It's an easy way for skilled professionals with the proper equipment to make copies or matching pieces.

♦ The detail and sharpness of the wax or resin models are usually better when made with a mill or 3D printer.

♦ After the costs of the CAD software and 3D printer or mill are paid, it is usually the least expensive way to produce a single piece or several models.

For more information on CAD, consult the February 2011 issue of *MJSA Journal*, entitled "Celebrating CAD: How software systems compare–and how users are benefitting."

Disadvantages of creating models with CAD & 3D printers or mills

♦ The equipment required for 3D printers and mills is very expensive.

♦ Training and experience is required to be able to create a good CAD model. Training is normally anywhere from a week to a month long, but the time it takes to get proficient with the CAD software can take a few or many months, depending on the individual.

♦ For many jewelry professionals, spending much of their time on a computer is not as pleasurable and fulfilling as hand fabrication and wax carving.

Advantages of creating jewelry directly with 3D printers or mills

♦ Large companies can save time and money by eliminating the casting process and thereby offer lower-priced jewelry and increase their profits.

♦ Porosity is reduced to none or almost none with a metal CD printer or mill.

♦ The uniformity and evenness of the settings and mountings are superior to those created by casting and hand fabrication

Disadvantages of bypassing the wax model production process and creating jewelry directly with 3D printers or mills

♦ Currently, it is only cost effective for large companies or jewelers who make high-end jewelry, or for jewelers who have the engineering know-how to make their own machine. Prices may come down in the future.

♦ Compared to hand fabrication, it's an impersonal way of creating jewelry. Some customers take pleasure in personally knowing the jeweler who created a piece with his or her own hands.

♦ For artisans and craftsmen, direct 3D metal printing and milling would not be a pleasurable way of creating jewelry. However, Doug Kerns of Gemvision notes "I have spoken to many jewelers who would consider themselves artisans and have invested in CAD and it has changed their lives as it has given them the ability to push the creative side of the design process. In other words, things that may have been impossible to do by hand carving or fabrication are now possible due to the CAD/CAM process."

Fabricating Custom Palladium Earrings

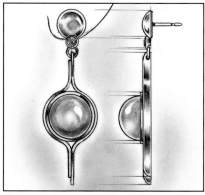

Fig. 7.34 Fabricating the metal components. *Photo by Mark Mann.*

Fig. 7.33 Custom earring design by Lainie
Mann with mabé pearls and pink sapphires.

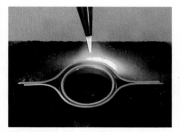

Fig. 7.35 Using graph paper to ensure identical forming. *Mark Mann.*

Fig. 7.36 Soldering the components
together. *Photo by Mark Mann.*

Fig. 7.37 Finished earrings by Mark Mann.

Hand Fabrication

Hand fabrication, which is also simply called **fabrication** is the oldest way of
making jewelry. The piece is entirely made with hand procedures such as bending,
carving, filing, folding, hammering, piercing, sawing and twisting. Other fabrication
techniques include:

Anticlastic raising: A metal forming process in which the center of a flat metal sheet
is compressed while its edges are stretched. The resulting form resembles a saddle and
has two curves at right angles to each other moving in opposite directions.

Filigree: a delicate design of fine wires soldered together in the frame of a heavier wire
or onto a flat base.

Forging: Shaping metal using hammers, anvils and heat.

Fusion: The uniting of two metals without solder by heating them to their melting
points. Also, the process of melting metal to produce interesting forms and textures.

Mokume gane: An ancient Japanese technique of making wood grain patterns using
layers of contrasting colored metals such as copper and yellow gold and/or silver.
Mokume gane, which means "wood-grain metal" in Japanese, was first used to decorate
samurai swords.

Fig. 7.38 Mokume gane ring by James Binnion made of silver, palladium white gold and 18K gold. *Photo by Hap Sakwa.*

Fig. 7.39 Mokume gane ring by Chris Ploof Designs made of silver, palladium white gold and 18K gold. *Photo by Robert Diamante.*

Fig. 7.40 Silver metal clay and titanium brooch by Holly Gage. *Photo: Holly Gage.*

Fig. 7.41 Art Clay silver ring with UV resin by Jackie Truty of Art Clay World, Inc. *Photo by F & W Publications.*

Precious Metal Clay (PMC) molding: creating metal objects with a moldable metal made of very small particles of metal mixed with an organic binder and water. After the "clay" is formed into the desired shape and dried, it can be carved, filed and even set with stones that can withstand high temperatures. Then the binder is burned away in a kiln, on a gas stove or with a hand-held torch, leaving a piece of jewelry or other metal object.

Precious Metal Clay was developed in the early 1990's by Japanese metallurgist Masaki Morikawa and patented by Mitsubishi Materials of Japan. **Art Clay** is another brand of moldable metal, which is produced by Aida Chemical Industries in Japan and distributed in the US by Art Clay World, USA.

Soldering: The process of uniting two pieces of heated metal together with **solder**, an alloy designed to melt at a lower temperature than the metal to be joined.

Welding: the joining of two metal pieces by using a laser, high pressure, electric current, heat + hammering, or heat + a filler with a high melting point.

Advantages of hand fabrication

♦ It usually requires less cleanup and finishing than casting. Therefore the backs of hand-fabricated pieces are often smoother and brighter than cast pieces.

♦ It allows exclusivity and great versatility. Each item can be unique.

♦ It shows the individuality of the craftsman.

♦ Hand-worked metal is usually stronger and denser than cast metal. One exception is Precious Metal Clay that has not been work hardened after heating.

♦ Hand-worked metal can be stronger and denser than cast metal.

♦ Hand-fabricated jewelry can be more lightweight than if it were cast, which makes it an ideal way to make comfortable earrings.

Disadvantages of hand fabrication

♦ It's usually a very time-consuming process, although in certain cases hand fabrication can be easier and faster than other methods.

♦ It's often more expensive, because of the extra time required.

Occasionally people neglect to consider purchasing hand-crafted jewelry because they assume it will be too costly. This is a mistake because hand-fabricated jewelry can often be affordable. Hand-crafted pieces offer buyers the privilege of owning an individually created piece rather than something that is mass produced.

Handmade Jewelry

Within the jewelry industry, the term "handmade" has a variety of meanings and connotations. According to the Federal Trade Commission of the United States, the term "handmade" should only be applied to jewelry which is made entirely by hand methods and tools. If any part of the piece is cast or die-struck, it is not handmade. (http://www.ftc.gov/bcp/guides/jewel-gd.shtm). Appraisers often use this meaning for the term "handmade."

Some jewelers describe custom-made jewelry as handmade if most of the work is done by hand. They reason that a cast piece made from a hand-carved wax and then finished and set by hand can require just as much creativity and work as a hand-fabricated piece.

Chain manufacturers make a distinction between chain which is handmade and machine-made. If a rope chain, for example, is assembled by hand, it is considered handmade even if the loops have been formed by a machine.

Since "handmade" is a more generic, less precise term than "hand-fabricated," goldsmiths in America generally prefer to have their hand-crafted pieces described as "hand-fabricated" or simply "fabricated." This way it's clear that no part of the piece is cast or die-struck. To a few people, "handmade" may even sometimes have negative connotations such as "homespun" or "unsophisticated."

Since the word has so many meanings, it's best to ask salespeople and jewelers to define what they mean when they use the term "handmade"; this will prevent misunderstandings.

Fig. 7.42 Cast 18K white gold frame with hand-fabricated wire design and setting. *Brooch and photo from The Roxx Limited.*

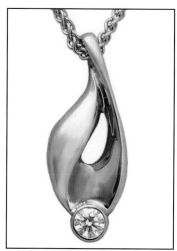

Fig. 7.43 Cast piece with a hand-fabricated bezel. *Pendant and photo by Hanna Cook-Wallace.*

Why Manufacturing Methods are Often Combined

Jewelers like to take advantage of the benefits of the various manufacturing techniques. For example, the diamonds of engagement rings are often put in die-struck or hand-fabricated prong settings. Since the metal of these settings is usually harder and denser, they can hold the diamonds more securely than those which are cast. Also, the shape and thickness of the prongs are generally more consistent. Another advantage of using die-struck settings is that the labor cost may be reduced.

In his book, *The Retail Jeweller's Guide* (p. 200), Kenneth Blakemore mentions that it's common practice in Britain to use a cast head and hand-fabricated shank to make a gem ring. "The drawn wire (hand-fabricated) shank has more elasticity than has a cast one, and will better stand up to the stretching entailed in sizing it."

Long parallel lines or rims on an item can be of a more consistent thickness if they're hand-fabricated rather than cast. It's often better to make bracelet and necklace catches by hand so they can be an integral part of the piece.

Figure 7.42 is an example of how a jeweler can combine methods to produce a high quality piece in a cost-effective way. The scroll-like frame was cast from a hand-carved wax model. A rubber mold of it was made so the design cost could be shared by other pieces cast from it. The wire design and setting were hand-fabricated because it was easier to bend the wire than to carve a wax mold and cast and finish it. In addition, the wire patterns are cleaner and more attractive. When other scroll frames are cast from the mold, the setting and wire design can be changed to give each piece a unique look.

Sometimes people regard cast and stamped jewelry as cheap. This is inaccurate. High-quality jewelry can be made with any of the manufacturing methods described in this chapter. What counts is that the jeweler is skilled and that the method(s) chosen suit the needs of the buyer and the piece.

8

Real or Fake?

Imagine that you've just inherited a large box of jewelry. You'd like to know if any of it is real gold, platinum, palladium or silver. This could be determined with a combination of the following tests:

Tests That Require No Acids

Magnet Test: Touch the magnet to the metal portion of the jewelry. If it picks up the jewelry, then the jewelry is not gold, silver, palladium or platinum; it is probably steel or iron. Metals that are strongly magnetic and that can become magnetized are technically referred to as **ferromagnetic**. Metals with a weak attraction to magnets are described as **paramagnetic**. For example, platinum and palladium are paramagnetic, but the attraction is not obvious or strong enough for them to be picked up by a magnet. Magnetism expert Kirk Feral says that you can detect a magnetic response in paramagnetic metals by placing the jewelry on a Styrofoam raft floating in a bowl of water and holding the magnet next to the metal portion. High-purity gold and silver jewelry will not be drawn toward the magnet, but platinum and palladium will be strongly pulled. If platinum is alloyed with cobalt, another ferromagnetic metal, the magnetic attraction is stronger, but still not as strong as that of steel, iron or pure cobalt. Metals that are repelled by a magnet such as gold are called **diamagnetic**. Table 8.1 shows the relative magnetic susceptibility of various metals based on information provided by the Fermi National Accelerator Laboratory (Fermilab) in Batavia, Illinois, at www-d0.fnal.gov.

Table 8.1 Magnetic Susceptibility (a plus sign indicates paramagnetism)

Aluminum	+16.5	Lead	-23	Ruthenium	+39
Bismuth	-280.1	Magnesium	+13.1	Silicon	-3.12
Cadmium	-19.7	Manganese	+511	Silver	-19.5
Cobalt	Ferro	Mercury	-33.5	Tin (gray)	-37.4
Copper	-5.6	Nickel	Ferro	Tin (white)	+?
Germanium	-11.6	Niobium	+208	Titanium	+151
Gold	-28	Palladium	+540	Tungsten	+53
Iron	Ferro	Platinum	+193	Zinc	-9.15
Iron Oxide	+7200	Rhodium	+102	Zirconium	+120

Source: http://www-d0.fnal.gov/hardware/cal/lvps_info/engineering/elementmagn.pdf

Magnets can be purchased at toy or hardware stores in various strengths. Neodymium magnets are very strong, but often too strong to be safe to use around children. Many refrigerator magnets are too weak to give reliable results for this test, but some will work.

Gemstones can also be paramagnetic or diamagnetic. For information on how to use magnetism to help identify gems, go to **www.gemstonemagnetism.com** by Kirk Feral. According to Feral, the magnetic wands needed for gem testing are not commercially available, but anyone can easily make a wand. Special neodymium magnets suitable for gem testing can be purchased online for just a few dollars at http://kjmagnetics.com/. The standardized size and strength for gem testing is a ½" x 2" (12.7 mm) cylinder, N-52 grade (product code D88-N52). Keep magnets away from electronic items.

Heaviness Test: Bounce the jewelry piece in your hand. If it feels unusually light, it's probably not gold and definitely not platinum. To learn how gold or platinum should feel, bounce some in one hand and then compare it to a piece of costume jewelry in the other hand. Keep in mind that hollow gold jewelry or chains may feel like imitation gold. Silver, palladium, 10K and 9K gold are similar in heft.

It will be easier to compare and identify metals by their weight if you know their **specific gravity** (their weight compared to the weight of an equal volume of water at 4°C.) For example, the specific gravity of pure gold is 19.36, meaning it is 19.36 times heavier than water at 4°C. Listed below are the specific gravities of some of the metals used to make fine and fashion jewelry

Table 8.2 Specific gravities of various metals and alloys

Iridium	22.4	Gold, 14K yellow	13.4	Jewelers bronze	8.7
Platinum	21.4	Palladium	12.2	Brass	8.5
Tungsten	19.3	Gold, 10K yellow	11.6	Iron	7.9
Gold, 24k (pure)	19.3	9K yellow	11.3	Stainless steel	7.8
22K yellow	17.3	Silver (fine)	10.6	Tin	7.3
18K white	15.7	Sterling	10.4	Zinc	7.1
18K yellow	15.5	Copper	8.9	Titanium	4.5
14K white	13.7	Nickel	8.8	Aluminum	2.7

Source of most data: *Jewelry: Fundamentals of Metalsmithing*, by Tim McCreight

Karat & Fineness Stamp Test: Look for a karat or fineness mark on the piece (figs. 8.1–8.4 and 1.8–1.14 in Chapter 1. Fineness is the amount of gold, platinum, palladium or silver in relation to 1000 parts, whereas **karat** value is the amount of gold in relation to 24 parts. Some marks are readable with the naked eye, but it's easier to decipher them with a 5- or 10-power hand magnifier. The karat or fineness stamp is only an indication of the gold, platinum, palladium or silver content, not proof. On the other hand, the lack of a mark doesn't necessarily mean the piece is not gold, platinum, palladium or silver.

Fig. 8.1 "PT 950" for platinum 950, "750" for 18K gold on a platinum and 18K gold ring. *Photo © by Renée Newman.*

Fig. 8.2 A "14 ct" mark for 14K gold. The British write "karat" as "carat." *Photo © by Renée Newman.*

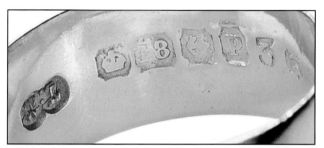

Fig. 8.3 British hallmark, left to right: the maker's mark "W.L," 18K gold marks, Birmingham Assay Office mark, "p" date letter for 1914; mark of "**36**." *Photo: The Three Graces Antique Jewelry.*

Fig. 8.4 Sterling silver "925" mark and the trademark. *Photo © by Renée Newman.*

Occasionally, 18K gold may appear to be stamped 10K and vice versa because the stamps may look similar.

Gold and platinum chains will typically have a karat or fineness stamp on the clasp and on a separate plate or link attached to the chain. Sometimes the clasp is plated, but the chain is gold or vice versa. The various components of gold earrings tend to have their own separate marks too. A professional-looking karat stamp on an earring post is an excellent indication of its gold content. It's not worth it for counterfeiters to buy the equipment needed to produce the tiny stamp on a post. They can make more money producing, for example, a fake 18K gold Italian necklace.

Trademark Test: Look for a manufacturer's mark along with the karat or fineness stamp. This is added assurance that the gold content is as stamped, and it is mandatory if the precious metal content is stamped on the piece. If there is no fineness stamp, a trademark is not required. Keep in mind that existing trademarks and hallmarks can be counterfeited and fake ones can be created. So even if the jewelry has a trademark, it should still be tested, especially if you're not familiar with the trademark and the type of jewelry that a particular manufacturer produces.

Color Test: Note the color of the jewelry. If it's yellow gold, it should normally not look like copper or brass. Check to see if the color is evenly distributed. If it isn't, gold plating may have worn off the base metal or the piece might be only partially gold. For example, the back of a necklace may be 18K gold and stamped as such, whereas the front, which is less likely to be tested, may be 10K, or another metal plated in gold.

Check around the hinges or clasps for color differences and signs of wear. This is an indication of plating (fig. 8.5). This can be verified by filing the piece in an inconspicuous spot and comparing the color underneath to the color on the surface.

The color test can sometimes be used to distinguish white gold from platinum or palladium. White gold often has a slight yellowish tint if it hasn't been plated with rhodium. Silver often has tarnished areas, unlike platinum and palladium which do not

Fig. 8.5 Note the gray areas where gold plating has worn away.

tarnish. Because silver and stainless steel also weigh much less than platinum, this factor should be taken into consideration along with color tests.

In general, the color test should only be used as an indication of a metal's possible identity, not as proof. Karat gold comes in a range of colors depending on the metals it's alloyed with. If, for example, it contains a lot of copper, it may be mistaken for copper. Brass can resemble some gold. Consequently, other tests should also be used in conjunction with the color test when identifying metals.

Closed Back Test: If the jewelry contains gemstones, look at the back of the setting. Is the bottom of the stone(s) blocked from view or enclosed in metal? Do the stones look like they have been glued in rather than set? Fake stones are commonly set or glued in fake-gold mountings. Genuine gems usually have open-back settings.

Price Test: If the jewelry is for sale, is it being sold at an unbelievably low price? If so, it's likely to be plated or underkarated. In the case of platinum, part of the piece may be white gold plated with rhodium. Even jewelers use price as a warning sign of misidentified metals. They know it's pointless to sell gold or platinum below cost or without making a profit. When a deal seems too good to be true, it probably is.

Torch Test: Jurgen Maerz, the technical consultant for the Platinum Guild, uses this test to confirm or disprove if a metal is platinum. Using a graver, he removes a tiny splinter of metal from a non-visible area. He places the splinter on a ceramic soldering surface and then attempts to melt it with a hot flame. If the metal forms a ball, then it's not platinum. If, on the other hand, it retains its shape and glows brightly like a miniature sun, then it's platinum or a platinum alloy.

If an unknown white metal loses its luster when heated and takes on a blue-violet surface oxide, the metal is possibly palladium. Platinum remains colorless and does not lose its luster when heated. One exception is cobalt platinum, which may oxidize, causing the platinum to lose its polished luster. Depending on the amount of heat used, the color may be a light blue or a light purple, but it's not as strong as that of palladium. Source: the *Palladium Technical Manual* by Johnson Matthey Precious Metals, UK Edition.

X-ray Fluorescence Test: This test requires an x-ray fluorescence instrument and is used to analyze the metal content by assay offices, refiners and other dealers who buy and sell precious metals. The device, which is referred to as an XRF analyzer or energy-dispersive x-ray fluorescence(EDXRF) spectrometer, is available as a portable hand-held device (fig. 8.6) and as a large desk-top or small portable enclosed system (fig. 8.7).

Fig. 8.6 EDXRF(Energy dispersive X-ray fluorescence) handheld ElvaX Prospector. *Photo courtesy Xcalibur XRF Service.*

Fig. 8.7 Olympus GoldXpert mobile XRF analyzer. *Photo courtesy Olympus NDT.*

XRF analyzers not only tell you if an object is gold, platinum, palladium or silver, they also indicate other alloying elements and the percentage of each metal. For example, it can determine if and how much nickel, lead or cadmium are present, and if so, how much. It can even determine the percentage of lead in toys, paint, or lead-glass filled rubies. Governmental agencies and many other industries use XRF spectroscopy for a variety of purposes such as mining, exploration, customs and border patrol inspections, environmental monitoring, poison detection, coating and wall thickness measurements, scrap metal sorting and recycling, analysis of drywall, soil, sediment and hazardous materials, etc.

XRF works by exposing a sample to a focused beam of X-rays. The atoms of the sample absorb energy from the X-rays, become temporarily excited and then emit secondary X-rays. Each chemical element emits X-rays at a unique energy. By measuring the intensity and characteristic energy of the emitted X-rays, an XRF device can provide both qualitative and quantitative analysis regarding the composition of the material being tested.

Handheld and portable enclosed devices typically sell for between $17,000 and $50,000; the enclosed devices offer more protection from X-ray exposure. Large machines for industrial analyzers can sell for $100,000 or more. You can also rent XRF analyzers for short-term use. They are easy to use; you simply place the sample in the machine and read the numbers. However the devices must be calibrated properly, or they may give inaccurate results. Fire assays are more accurate than XRF devices, but not as convenient. The refiner, Umicore, estimates on its website that differences between the two techniques can be up to 2% (e.g., 48% instead of 50% gold content), depending on the calibration of the equipment and the nature of the material. Filing is required to verify that a piece is not plated or bonded with another metal. In addition it is wise to test the piece in more than one position because the metal may not be the same throughout.

The Birmingham Assay Office says that XRF testing with the right machines and highly sophisticated computer programs can be extremely accurate. In fact, their XRF testing results have been so consistent that since 2005, they have moved to XRF testing for hallmarking jewelry. Currently more than 98% of their hallmarking is done with XRF machines and specialized programs for each fineness mark. The results are regularly checked with assaying techniques.

Many people use multiple tests for analyzing metals. The touchstone test described in the next chapter is still widely used by professionals for determining gold content.

Nitric Acid Testing for the Layperson & Hobbyist

Acid testing is relatively inexpensive. You can separate imitation gold from 14K yellow gold and above with the aid of nitric acid or just one bottle of 14K gold testing liquid, which is diluted nitric acid. (At a jewelry supply store or rock shop, it will normally range from $5–$10. An acid-resistant plastic bottle is better than the glass type because it's non breakable, airtight and spill-proof.) For safety reasons you should also have some baking soda, rubber gloves (the kitchen type will do) and safety glasses (these cost as little as $5 in jewelry supply, hardware and chemical supply stores). For accurate testing, the acid solutions should be no more than six months old.

After you have the above supplies, as well as a jar of water and some white paper towels, follow the steps below. It's advisable to first practice on known metals.

♦ Take your jewelry and testing supplies to a well-ventilated area (e.g., outdoors, in a large garage or on a patio).

♦ With the rubber gloves and safety glasses on, place a small drop of acid on an inconspicuous spot of the piece you want to test. If the drop bubbles and hisses, the piece is definitely not gold or platinum. (If the piece you're testing is 10K gold or less, the acid drop will probably leave a brown stain. To avoid staining the piece, use the touchstone test described in the next chapter.)

♦ Observe if the drop changes color. The list of metals below with their color reaction to nitric acid or 14K solution will help you interpret the results.

18K gold or higher	no change
14K gold	no change or a slight browning
9K & 10K gold	turn brown
low karat gold	turns brown and green
platinum	no change
rhodium	no change
silver	turns white or gray
stainless steel	usually no change
copper, nickel	turn green
brass	turns green
palladium	turns a golden yellow

If you question the color change, also blot the piece with the white paper towel and look at the stain. A green, yellow and/or brown color is often more evident as a stain. Keep a mental note of the color of the stain of various metal types. This is also a means of identifying metals.

Note that white metals other than white gold may show no reaction to the acid. The heaviness of platinum and lightness of stainless steel are helpful in separating these metals from white gold.

Pure yellow gold is the only yellow metal which shows no visible reaction to nitric acid, although nitric acid can dissolve a minuscule amount of gold. If there is no color change or bubbling, this means that the yellow piece is at least 14K gold or gold plated. (10K gold turns brown due to the high percentage of alloy metals.)

Areas where plating has worn away may become more obvious when a drop of acid is placed on them. For example, yellow **vermeil**, which is sterling silver plated or layered with gold, may show noticeable gray areas when exposed to nitric acid. (fig. 8.8). (**Sterling silver** is 92.5% silver and 7.5% copper and/or other elements. Its fineness mark is **925**.)

Fig. 8.8 Gray stains from nitric acid on a filed, gold-plated sterling silver chain.

Continue with the next steps.

♦ If the drop of acid shows no reaction, file the piece in an inconspicuous spot. Look at the spot with a hand magnifier and compare the color of the metal surface to that of the filed area. If they are different, the piece is probably plated.

Karat gold is sometimes plated to give it a stronger yellow color. For example, many 14K Italian chains are plated with 18K gold. Therefore, color differences are not always an indication of imitation gold.

♦ After the piece is filed, place a drop of acid in the notch. Look for bubbling and for green, brown, white or gray colors. If you see none, look at the piece close up through a hand magnifier with your safety glasses on. (The fumes from the acid can damage your eyes.) White gold plating on sterling is one of the hardest imitations to detect and often magnification is required to see its reaction to the acid.

Gold buyers report that the plating on jewelry is getting thicker, particularly on the areas where the piece is most likely to be tested (e.g., on clasps, between links and on the inside of bracelets). Even jewelers have been fooled by some of the plated bangle bracelets and heavy link chain that are currently being sold. Filing is a necessity, when testing gold.

♦ After testing with acid, dip the piece in a jar of water mixed with a couple of spoonfuls of baking soda. This will stop the acid reaction on the alloy of the metal. Then rinse the piece with water. If acid gets on your rubber gloves, you can dip them in the jar and rinse with water. The acid must be neutralized, not merely diluted, especially if it gets on your skin.

♦ Put any paper towels used for blotting or wiping acid in a ziploc plastic bag. Add some baking powder to the bag. If the towels have more than a few tiny droplets of acid on them, baking soda may not be enough to neutralize the acid. See the section on acid disposal for proper disposal of hazardous wastes.

Agua regia—nitric + hydrochloric acid—is helpful for testing 18K gold and above. The next chapter explains how it is used with the touchstone test.

Figs. 8.9 Metal sample results viewed close-up. The silver testing drop looks orange on the stainless steel spoon, red on the paper towel below the ring in the center, brownish red inside the center ring, but the drop shows no color change on the 14K white gold mounting or on the paper towel below it. (Actual colors may change during printing, Do the experiment yourself to see the true results.) *Photo © by Renée Newman.*

Silver Testing

Sterling silver can be detected by placing a drop of potassium dichromate solution on the test metal surface. Jewelry supply stores sell it in small plastic bottles under the name of "Silver Testing Solution" for about $5–$10 a bottle. The liquid turns red and leaves a red stain on a white paper towel with it comes in contact with sterling silver. After it dries, it may look gray. Figure 8.9 shows how a sterling silver ring reacts to the silver testing solution in comparison to a stainless steel fork and a 14K white gold mounting. When wet, the solution looks brownish red on the sterling silver and bright red on the white paper towel below it; the drop looks orange on the stainless steel; but the drop shows no color change on the 14K white gold or on the paper towel below the14K white gold. It is best to try testing metal yourself on known samples to learn how each metal reacts to the test solutions. Printed colors in books may be misleading.

Even though the drop did not affect the stainless steel or 14K white gold, it etched the surface of the sterling silver leaving a white mark instead of a shiny silver surface. Therefore this test should either be done on an inconspicuous spot on a test piece, or else be done on a touchstone, where the metal is rubbed onto the touchstone and the silver solution drop is place on the streak on the touchstone. See Chapter 9.

Palladium Testing

Sources differ as to how to test for palladium. Some recommend placing a drop of lab-grade iodine on a cleaned surface of the white metal to be tested. If the iodine turns black, then the metal is said to be palladium. The problem with this test is that is doesn't always work. Some brands of iodine remain colorless on palladium and test like platinum.

Concentrated nitric acid or 9K, 10K and 14K solutions, which consist of diluted nitric acid, are more reliable liquids for testing palladium. Dippal Manchanda MSc CSci CChem FRSC, technical director of the laboratory at the Birmingham Assay Office in the UK, says that concentrated nitric acid consistently produces a golden yellow stain. I tested

950 palladium in Los Angeles with a newly purchased bottle of 14K solution and got the same result. The golden stain is especially noticeable when you blot the liquid drop on the palladium with a white paper towel.

Manchanda determines if an 18K white gold alloy is a palladium alloy by applying a drop of aqua regia to the metal and soaking the test acid (aqua regia) with blotting paper. When the reaction is complete, he applies Dimethyl P-Nitroso Aniline to the test paper. If palladium is present, a red color stain appears on the test paper.

Acid Testing for Jewelry Trade Professionals

The acid testing procedures above also apply to professionals, such as jewelers, appraisers, pawnbrokers, etc. However, greater precision is needed and legal requirements can differ.

In some communities, it's illegal for businesses to buy or own any amount of acid without first obtaining a permit. (This includes the small bottles of gold testing liquid sold at jewelry supply stores.) According to a Pasadena, California fire department official, for example, businesses with small amounts of acid are often the least likely to know the proper precautions for using and disposing of acids. Small amounts of acids in many bottles can also add up to large amounts. Consequently, Pasadena requires businesses to have a permit in order to buy or use any amount of acid. Households and hobbyists, however, are normally exempt from this regulation. Some other communities require permits only for large quantities of acid. Before buying or using acids, check with your local city government and/or fire department to find out what the regulations in your area are.

Jewelry-trade professionals are better off using pure acids rather than the premixed acids sold in jewelry stores. Some reasons for this are:

♦ Mixed acids tend to decompose faster than single acids. Gold testing liquids sold in the United States normally contain varying amounts of both nitric and hydrochloric acid (HCl), also called **muriatic acid**. Even the 10K liquid has a little HCl in it, even though jewelers typically test 10K gold only with nitric acid.

♦ The strength of the premixed acids may vary from one batch or supplier to another. For consistent results, the same strength and type of acid should always be used for a specific test. You can control this better if you buy the individual acids yourself.

♦ It's easier to tell when pure acids are fresh because they fume more than the premixed gold testing liquids. A lack of fumes indicates the acid has lost its original strength. Stale acid can make 10K gold appear to be 14K.

♦ Pure acids are stronger than the ready-mixed acids. Consequently they may show clearer reactions. For example, the darker gray color reaction of pure nitric acid on silver is easier to see than the lighter gray produced by a 14K premixed solution.

Pure acids can be bought at chemical supply stores. They don't need to be diluted with water for gold testing. In fact the stronger they are, the more obvious the results. (Nitric acid from most chemical supply houses in Southern California is about 70% strength and HCl about 30%). Instead of using 14K gold testing liquid, professionals need only apply a drop of straight nitric acid to the unknown filed metal to determine if it is gold. HCl is needed to determine the fineness of high-carat gold and to test for platinum. The next chapter explains that test.

Since the bottles from chemical supply stores contain more and stronger acid (4 oz may be the minimum size), greater precautions are needed when handling them. If you're not accustomed to pouring acids from one bottle into smaller bottles for testing, either have an experienced person do this for you or ask the chemical supply house to explain how. They will probably recommend stronger rubber gloves and a plastic apron. The small acid-resistant plastic bottles convenient for gold testing might not be sold at the chemical store. Jewelry supply stores often have them. When acids are exposed to air a lot, they get stale faster. Therefore, it's best to use the smaller bottles for gold testing and add fresh acid when needed.

Be careful where you store the acids. Besides being highly toxic, their fumes can damage your computers, appliances, and anything else you have that contains copper, nickel, brass or silver. Naturally, make certain children can't reach them.

Disposing of Acids

Regulations and arrangements for disposing of hazardous wastes vary from one town to another. Some cities may have a place where you can drop off your bottles of old acid at any time free of charge. Other cities may arrange a drop-off point at intervals such as once a week to twice a year. Depending on the type and quantity of acid, city officials may tell you how to neutralize it. A stronger substance than household baking soda is usually required.

To find out your city's disposal arrangements, call the fire department, city hall or Board of Public Works. They can direct you to the department in charge. Cities usually provide households with a free means of hazardous waste disposal.

Businesses may have to pay for acid disposal if it's not included in the price of their permit or license to use acids. Local regulations should be checked.

If you need to quickly dispose of some gold-testing acid but your city's disposal services are inconvenient, ask your jeweler if they or someone they know can take it from you. Some will accept it, just as some full-service gas stations will take old motor oil.

Never dispose of acids in garbage cans. Pets, children and food scavengers can be permanently damaged if they come in contact with the acid while rummaging through the trash.

Safety Tips for Using Acids

This chapter has already indicated precautions for acid testing. It helps, however, to have them summarized.

♦ Use acids only in a well-ventilated area. Don't breathe in the fumes.

♦ Wear rubber gloves.

♦ Wear safety glasses, especially when opening or pouring acids and when looking at acid reactions with hand magnifiers.

♦ Open acid bottles away from your face. The fumes may rush to your face.

♦ Never pour water into acid. This makes acid spit. To mix the two liquids, pour acid gradually into water.

Fig. 8.10 Sign outside of a store in Italy. *Photo by Debra Sawatzky of Value the Past Appraisals.*

♦ Keep baking soda nearby to neutralize acid.

♦ Do not let acid flow onto gemstones. Acid will etch or discolor many gems.

♦ Keep acid bottles tightly closed. Sometimes the acid eats away at the lids, and they need to be replaced to prevent the fumes from escaping.

♦ Store acid in a safe place away from children. Avoid extreme temperatures.

The Consequences of Buying Counterfeit Goods

In Europe, it is illegal to own counterfeit items such as fake Rolex watches and fake Gucci bags, but fines vary from one country to another. Anyone caught with counterfeit goods in Italy, for example, could be fined up to 10,000 Euros; some tourists have already learned that getting caught by the Italian police may cost more than having their wallet snatched by pickpockets. In France, fines can reach up to 300,000 Euros.

In North America it is generally legal to own a fake but not to sell one; however, this could change. In 2012, new legislation was introduced in New York City that would make it a misdemeanor to buy knockoffs. Punishment could range from up to a $1000 fine to as much as a year in jail.

If you get caught in Customs with counterfeit goods, expect to have the goods seized if they are discovered and recognized as fakes. A seizure report could be added to your computerized record, making customs searches more likely in the future.

Below are some tips to help you have positive buying experiences and avoid purchasing misrepresented products.

Buying Tips for Branded Items

- **Save all receipts**. Get detailed written receipts and e-receipts if possible. It's easier to insure and resell well-documented goods. Documentation can also add value to estate pieces; prove ownership; and prevent you from being falsely accused of theft.

- **Pay with credit cards rather than cash**. This is further proof of ownership and offers you protection if the product is defective or misrepresented.

- **Shop in your home country for legitimate branded items** such as European watches in order to avoid servicing or misrepresentation problems.

- **Declare all items purchased abroad**. Besides being the right thing to do, you may be unable to collect insurance or legitimately resell items from abroad if required customs duties have not been paid.

- **Buy from knowledgeable and reputable sellers** who verify the origin of their merchandise. This is especially important when buying goods second hand. Otherwise you could end up with stolen merchandise, knockoffs or so-called antiques that are not really antiques.

- **Have insurance appraisals done by a legitimate, competent appraiser who expects proof of purchase on designer or branded items**. There are two ways to prove that an item is an authentic designer or branded piece worthy of added value: with a detailed receipt or by sending it for verification to the designer or company that made the piece. The latter is usually inconvenient and costs money. Unfortunately, you can't simply rely on trademark stamps because they can be faked.

The above information is from Debra Sawatzky, an appraiser specializing in antique and estate jewelry. A list of independent gem and jewelry appraisers and appraisal organizations is available at **www.reneenewman.com/appraisers.htm**. For information specifically on antique and estate jewelry and personal property, go to **www.valuethepast.com**.

9

Determining Karat Value & Fineness

To determine the value of precious metal jewelry, one must determine the amount of gold, platinum, palladium or silver in the piece, in other words its fineness (or karat value in the case of gold). The easiest way is to look for a stamp on the piece next to a trademark. However the stamp is not always correct and can be counterfeited. Chapters 2–5 discuss and show fineness markings on gold, platinum, palladium and silver. Gold fineness markings can be verified with the touchstone test described below:

Touchstone Test

To do this test, you'll need gold of known karat value and a **touchstone**—a hard, fine-textured black stone sold in jewelry supply stores. Ideally it is made of basanite, a form of black jasper. Gold test needles, which are marked according to the karat value of their gold tips, are typically used as a comparison reference for gold. To test for 14K and below, nitric acid (or 14K fluid) is required. To test for 18K gold and above or or platinum, you'll also need some hydrochloric acid (HCl) or some platinum and 18K gold testing liquid. Follow the procedure below using the same safety precautions outlined in the preceding chapter, entitled "Real or Fake."

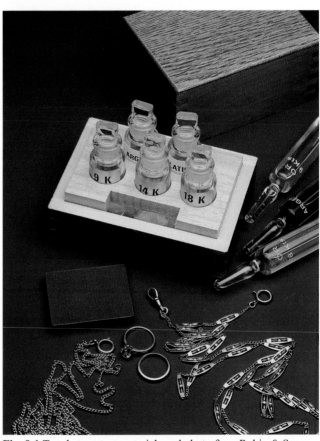

Fig. 9.1 Touchstone test materials and photo from Rubin & Son.

♦ Make a wide line on the touchstone by rubbing the unknown metal back and forth. Rub hard to reach the possible base metal. Filing may be necessary to detect thick platings.

Fig. 9.2 Results of an 18K white gold pendant tested on a touchstone with 10K, 14K, 18K and silver testing fluids. The 18K gold streak on the touchstone only disappears at the point where the 18K white gold fluid was used. None of the fluids changes the color of the metal streaks. The pinkish color of the silver solution drop, is the color of the solution. If the metal were silver, the drop would look red to brownish red. *Pendant mounting from Hubert; photo by Josian Gattermeier.*

Choose the needle that you think might have the same gold content as the unknown metal. With the needle, make a line of the same width and intensity either beside or above the unknown line. Compare their colors. Yellow gold usually has a deeper yellow color than metals that imitate it. If you wish, make another line with a different test needle.

♦ To test 14K gold, apply nitric acid or 14K liquid across the lines. Observe the color and check if they disappear. 14K yellow gold should remain yellow and visible. Occasionally there may be a very slight browning. 10K gold can be compared to a streak from the 10K gold test needle. The acid may turn it brown, but sometimes the streak disappears before you see the brown color.

If the metal is white and a nitric acid drop causes the streak to dissolve, this indicates a silver, nickel or tin alloy. If the streak is just darkened, this indicates a 9K or 10K white gold. A streak unchanged by nitric acid indicates that the metal could possibly be platinum, stainless steel, an aluminum alloy or white gold that's higher than 10K.

♦ To test 18K gold, clean the touchstone or use the opposite side of the touchstone. Add a drop of 18K testing solution, or add a drop of HCl after you apply nitric acid to the known and unknown streaks. (The mixture of these two nitric and HCl acids is called **agua regia**.

Unlike nitric acid, agua regia can visibly attack high-karat gold. Note the rate and degree of disappearance. 18K and 22K gold should react like the 18K and 22K test needle streaks. According to *The Retail Jeweler's Guide: Fifth Edition* (p 394), if the metal is white and the streak remains unchanged after HCl is added to the nitric acid drop, then:

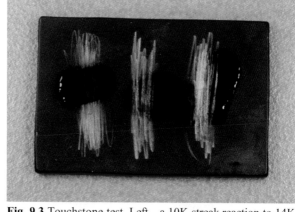

1. A white precipitate in the liquid indicates a silver alloy.
2. A darkened streak that dissolves indicates 9K or 10K gold.

Fig. 9.3 Touchstone test. Left—a 10K streak reaction to 14K liquid. The right streak appears to be of a higher karat value than the center streak. In this case, it's because the streaks were not applied with equal intensity. Both streaks were made with the same 14K gold piece.

3. A streak which slowly dissolves on the addition of HCl indicates a higher gold than 9K, a palladium jewelry alloy of 950 fineness, or a stainless steel.
4. A streak which remains unchanged on the addition of HCl indicates a platinum jewelry alloy of 950 fineness or 22K gold. According to Dippal Manchanda at the Birmingham Assay Office, concentrated aqua regia does not react with 22 ct gold; therefore they add potassium iodide to aqua regia, which produces enough iodide ions to react with gold to form gold iodide which is brown in color.

Some gold testing specialists recommend that you not test 14K gold on the same touchstone surface as 18K because the residue left by the HCl can speed nitric acid reactions and make 14K gold appear to be 10K. It's a good idea to mark which side is for testing 18K gold and above. Other specialists simply clean the touchstone surface thoroughly after each test.

♦ If the reaction is too fast for easy comparison, agua regia diluted with distilled water can be used. (But remember to pour acids into water, not vice versa, and do it slowly.) Nitric acid diluted with distilled water is used by some people to test 10K gold. As with the diluted agua regia, the reaction on the touchstone will be slower and probably easier to compare to 8K and 12K.

Michael Elliott of North American Metals in Van Nuys, California has found an easier method. He just rubs a little baby oil on the touchstone before the test. This slows the reaction of the nitric acid and HCl droplet mixture.

♦ In summary, use concentrated nitric acid to test 14K gold and below. To test gold over 14K, clean the touchstone or turn it over. Place a drop of nitric acid on the streak. Add a drop of HCl. Or place a drop of 18K gold testing fluid on the streak. Note the color and speed of the reaction. Try to match them to a known streak.

When testing rings, test both the top and bottom of the ring. It's not uncommon for the shank to be of a different fineness. It's also a good idea to test both the front and back of necklaces.

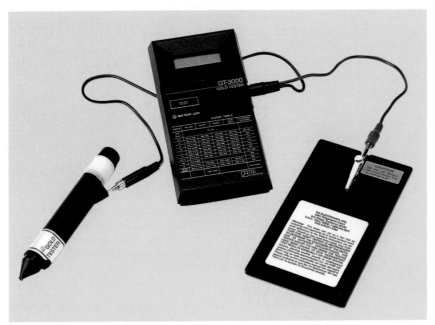

Fig. 9.4 TRI Electronics Electronic Gold Tester. *Photo courtesy Stuller.*

At first, practice the touchstone test on jewelry you know. The more you do the touchstone test, the more accurate your results from it will be.

♦ Clean the touchstone with wet and dry sandpaper and/or a paper towel. Afterwards, oil the touchstone surface with an oil such as olive oil or baby oil to protect it.

According to the London, UK Assay Office at the Goldsmith's Hall, the touchstone test is a more accurate means of testing gold than using an electronic gold tester. When done properly by an experienced tester, they estimate the touchstone test has an accuracy of at least 98%.

Electronic Gold & Platinum Testers

Portable electronic testers that measure the karat value of gold are available at jewelry supply stores (figs. 9.4 & 9.5). When the sensor tip of the tester with testing solution is applied to a metal, a digital reading will appear saying, for example, "18K" or "not gold." These testers range in cost from about $200 to $700. The cheaper ones are limited in the type of gold they can test and may only be able to indicate a 3-digit number for the gold content, instead of the direct reading of the karat value. A reference table explains the significance of the number.

The more expensive electronic testers can test for platinum and high karat 20-24K gold. They can identify platinum but are unable to indicate its fineness. Beware that certain types of stainless steel may test like platinum both on an electronic tester as well as on the touchstone test. However platinum is softer than stainless steel, so it can be distinguished from that metal by the ease at which it can scratch a touchstone. It's easier to form white lines on the touchstone with platinum and white gold than with stainless steel. Stainless steel also has a much lower density and heft.

If a white gold alloy contains palladium, an electronic tester may indicate that it is as much as two karats higher than its actual gold karat value.

Before testing an unknown material with a gold tester, you should try it on some metal with known gold content. Just as acids can go bad, so can the solution or gel used by the gold tester. Verify with the manufacturer that the chemical is within the expiration date. For accurate detection of gold filled or plated items, you must file the metals as with touchstone testing. Some testers may require continual calibration, whereas the TRI-electronics brand does not.

Even though gold testers are not an essential tool for jewelers, they do provide a quick, convenient means of checking gold fineness. To get the most accurate results, it is recommended that you carefully follow the user manual instructions and learn how to properly operate an electronic gold tester.

Two Test Examples

Specific examples provide a better understanding of the principles of gold testing. The first case is a braided herringbone bracelet (figs. 9.6 and 9.7). It feels lightweight, it's trademarked, and it has a fineness mark of 925.

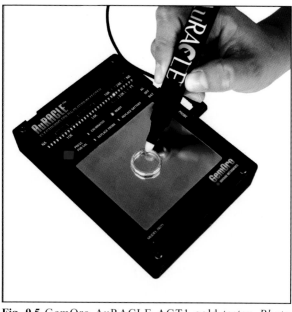

Fig. 9.5 GemOro AuRACLE AGT1 gold tester. *Photo courtesy GemOro and Rio Grande.*

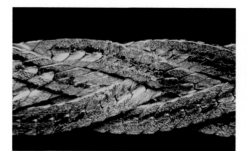

Fig. 9.6 Dark gray stains from nitric acid on a filed, gold-plated sterling silver bracelet. *Photo © Renée Newman.*

Fig. 9.7 Sterling silver mark, "925," on the clasp of the herringbone bracelet in fig. 9.6. *Photo © Renée Newman.*

The number "925" could be a stock number, but it is the fineness mark for sterling silver. When the piece is filed and viewed through a hand magnifier, the silver color underneath is easy to see. Nitric acid turns the metal underneath a dull gray, so it is in fact silver. If the bracelet is just rubbed lightly onto a touchstone, it tests as 18K gold. This bracelet is sterling silver with a layer of mechanically bonded 18K gold. It may sound like a very valuable piece, but its retail cost was less than $20. This first gold-testing example is an easy one; the next one isn't.

A neighbor brought the author a hollow gold bead necklace marked 14K (fig. 9.7). It had been passed down in her family and she wondered if it really was 14K. Since the beads were hollow, the weight was mis-leading. The necklace had the following characteristics.

Fig. 9.7 14K gold beads with a lighter base color than the surface. *Photo © Renée Newman.*

♦ It was not magnetic. If it had been, this would have indicated a core of iron or stainless steel, or cobalt platinum if the chain had been denser and heavier.

♦ There was no trademark.

♦ The color and surface texture looked fake, especially under magnification.

♦ The area between the beads was a different color than the outer part—light yellow with occasional black spots as opposed to deep yellow.

♦ When filed, the metal underneath showed a different lighter yellow color.

♦ The string holding the beads had a slight greenish tint.

♦ The reaction to the touchstone test varied from 14K to 18K depending on how hard the pieced was rubbed onto the stone.

The author did not put acid on the piece because it might have damaged the string between the beads. Since the test results were questionable, the necklace was taken to a gold assayer. Using the acid drop and touchstone test, he determined the necklace and clasp to be 14K gold.

To be absolutely certain the necklace was 14K throughout, one would have to cut through a bead and test the inner part with acid, but there was no point in doing that in this case. There's also a possibility that some beads are gold and some aren't.

Color differences between the outer surface and inner part of a gold piece do not necessarily indicate it is fake. 14K Italian gold chains are commonly plated with 18K to give them a deeper yellow color. 14K jewelry may even be plated with 24K gold.

The lesson from this example is that gold testing is not always clear-cut. Some pieces are easy to test, others may require professional assistance. But even experts admit that in some cases they have to cut completely through a piece to really know what's underneath the surface. Testing white metals, in particular, can be tricky.

To get an exact determination of gold content, a **fire assay** test must be done. A small sample of the gold in question is weighed and melted. Then by processes of oxidation, absorption, and dissolving in acid, the metals of the alloy are separated. The weight of the pure gold that results is compared to the original weight of the gold sample, and the fineness is calculated.

Fire assays are lengthy, expensive and destructive. So despite their accuracy, they aren't practical for general jewelry appraisal. As a consequence, the touchstone and acid-drop methods remain the most popular low-cost tests for gold and platinum. XRF devices are very expensive, but are non destructive and often used now to determine the identity and fineness of precious metals.

Why Precious Metals Tests Sometimes Fail

Precious metal testing must be done properly to be reasonably accurate. Even a metal assay can give false results if done incorrectly. When doing the acid tests described in this book, you should allow for a tolerance of about plus or minus two karats. However, if a metal is misidentified or the karat-value test results are off by four or more karats, this is probably due to one or more of the following reasons:

1. The metal being tested was not filed. This is essential for the detection of heavy plating and overlay.

2. The acid is old and stale. Weak acid can make 10K gold test like 14K. Check acids and electronic gold testers with known samples before using them to test unknowns.

3. The acid is contaminated. Do not let the acid or bottle tip come in contact with metal, the touchstone or other acid.

4. The test needles are corroded from acid fumes and therefore give false results. Do not store them next to acids.

5. The same side of the touchstone was used to test both high and low karat gold. The residue left by agua regia or 18K liquid during touchstone testing can make 14K gold test like 10K. Reserve a different side or stone for each test fluid used.

6. The width and intensity of the comparison streaks are different. A lightly rubbed streak can appear to be of a lower karat value than if it's rubbed hard on the touchstone (fig. 9.3).

8. Only one section of a piece was tested. Both the head and shank of rings should be tested. Necklaces and bracelets should be tested both on the clasp and at least one place in between. Keep in mind too, that a solder joint may test lower than the rest of the piece.

9. The person doing the testing was influenced by a karat marking or by what he or she wanted to see. Preconceptions can keep us from being objective and from doing the required number of tests. One advantage of the electronic gold tester is that if it says 14K, it's hard to read 18K. It's easier to misread the reaction of a streak on a touchstone.

10 Only one test was done. Gold experts often find it necessary to do more than one type of test. Weight, color, markings and magnetic attraction must also be considered when determining the identity of a metal. Gemologists rely on a combination of tests to identify gems. People testing metals should do the same.

Subtle Deception Techniques

There are legal ways to mislead buyers. For example, one gold dealer reports that in his town, merchants who had been convicted of selling 10K gold stamped 14K, discovered they could sell just as much 10K gold by stamping it **417**, the European fineness mark for 10K. Uninformed American buyers either think it's "14" written backwards or else a stock number. In most cases, they don't even bother to look at the stamp or ask what karat gold they are buying. The only thing that matters is the price tag. American-made 14K gold, however, is usually stamped 14K not 585.

Some sellers like to take advantage of the fact that gold prices can be based on grams or pennyweights (1 g = 0.643 pennyweight). They may sell by the gram and buy by the pennyweight. When getting price quotes by weight, be sure you understand the unit of weight used.

Hollow gold jewelry and chains offer consumers a big look at a low price. They also offer some sellers a convenient means of deceiving their customers. A hollow chain that looks double its weight may seem to be a bargain compared to a solid one.

It's not uncommon to see yellow gold-plated sterling stamped **925**. There are probably some sellers hoping it will be interpreted as a high purity of gold. When you see 925, think silver not gold.

Don't assume that all the jewelry in a 14K display case is 14K gold. 10K pieces may also be mixed in (and so may some 18K). Therefore, it's best to verify with the salesperson the karat value of the jewelry you plan to buy.

In addition to displaying 10K, 14K and 18K gold together, stores may also place jewelry with synthetic (lab-grown or created) stones next to pieces with stones of natural origin. There may be no signs saying lab-grown, and the salesperson may forget to mention they're man-made. The main reason for displaying lab-grown stones with mined ones is to make man-made stones appear as valuable when in fact they aren't. Limited space can also be a factor. Do not assume that all the gems you see in a jewelry case are of natural origin.

To avoid being misled, buyers must be informed. They should ask the following questions about the piece they wish to buy and have the answers written on their purchase receipt.

♦ What is the identity of the metal?
♦ If it is a precious metal, what is the fineness or karatage of the metal?
♦ Is the piece solid or hollow?
♦ What is the identity of the gemstones?
♦ Are the gemstones lab-created (synthetic)?

A reputable jeweler will include this information on sales receipts and appraisal reports whether you ask for it or not. He/she wants you to know even more than this about your jewelry because as one Chicago jeweler states, "An informed consumer is the Industry's best advertisement for expanded business. The more you know, the more you will feel confident in buying and the more you will appreciate the art and beauty of jewelry."

10

Iron, Stainless Steel & Tungsten

Iron is the most abundant metal on earth, but it is seldom found in the ground because it tends to oxidize and turn into rust. It is usually obtained from minerals such as hematite or magnetite, or from meteorites. Its chemical symbol, "Fe," was derived from the Latin word for iron: *ferrum*. The high strength, wide availability and low cost of iron alloys make them the most important construction metal for buildings and bridges.

Iron metal was used for tools and weapons probably as far back as 2500 BC, but it wasn't until about 1200 BC that it began replacing copper alloys, which have a lower melting temperature. Like pure copper, pure iron is soft, but can be selectively hardened by adding carbon and other elements during the smelting process. Iron was introduced into England about 1500; the first ironworks in America were established in Virginia in 1619.

Two forms of iron produced by smelting are cast iron and wrought iron. The *Encyclopedia Britannica* defines **cast iron** as an alloy of iron that contains 2–4 percent carbon, along with varying amounts of silicon and manganese and traces of impurities such as sulfur and phosphorus. **Wrought iron** is smelted more slowly and usually contains less than 0.1 percent carbon and 1 or 2 percent slag (residue from the smelting process). Tough, malleable and easily welded, wrought iron is superior for most purposes to cast iron, which is overly hard and brittle because of its high carbon content. Consequently, in the 1800's wrought iron replaced cast iron for construction. By 1890, wrought iron had largely been replaced by steel, a harder and stronger iron alloy. Today, wrought iron is primarily used for decorative purposes such as gates, railings, statues, and wall hangings. Thanks to the rising cost of gold, platinum and silver, wrought iron jewelry is making a comeback. It is sometimes combined with gold and other metals (figs. 10.4 & 10.5).

The definition of "iron" versus "steel" varies depending on the user and the context. Metalsmith Chris Nelson of Urban-Armour says,

> I use low carbon steel which is known as 1018. It's very malleable and does not harden much when cold forged or worked cold. Some metalsmiths use pure iron which is referred to as being quite "squishy", but there are very few suppliers for it in the U.S. I call my work iron jewelry, although it could be referred to as steel jewelry which I think would give it a more techno-commercial sound that I avoid using. The steel alloy I use is 92% iron with traces of other elements added, but if 14 karat (58% pure gold) can be called a gold alloy, then I choose to call a metal with 92% iron an iron alloy.

Besides being utilized for ornamental items and the construction of buildings and bridges, iron is used to make magnets, nails, pipes, red dyes, barriers against ionizing radiation, and nutritional supplements.

Fig. 10.1 Iron, pearl and diamond ring by Marsh & Co. *Photo courtesy Lang Antique & Estate Jewelry.*

Fig. 10.2 Forged iron & ocean jasper pendant. *Jewelry & photo by David Anderson of Erik Jewelers, Tonawanda, NY.*

Fig. 10.3 Berlin iron bracelet. *Jewelry and photo courtesy The Three Graces Antique Jewelry.*

Fig. 10.4 Forged iron enhanced with 18K & 22K gold. *Pendant & photo by Chris Nelson.*

Fig. 10.5 Forged iron enhanced with gold and shibuichi (pink silver—silver + about 75% copper. *Cuff & photo by Chris Nelson.*

Fig. 10.6 Machined bracelets made of grade 316 stainless steel. *Cuffs and photo by Pat Pruitt.*

Iron forms an important part of many proteins and enzymes that help the body function efficiently. The largest amount of iron is found in hemoglobin, a protein which gives blood its red color and which carries oxygen throughout the body. Sufficient amounts of iron in the body can help eliminate fatigue and muscle weakness and also benefits the immune system. Too much iron, however, can be toxic and lead to cell aging, cancer and heart disease.

Stainless Steel

Stainless steel watches have played a major role in the growth of stainless steel jewelry. Some men who would never consider wearing a gold or platinum bracelet, will buy one of steel to go with their watch. Men will also accessorize their watches with stainless steel pendants, wedding rings and necklaces. Women also buy stainless steel jewelry, particularly in Europe, where they are the primary buyers.

Steel is an iron alloy containing about 0.5 to 1.5 percent carbon and trace elements such as manganese and silicon, but as previously mentioned, steel alloys are sometimes simply alloyed iron. **Stainless steel** is steel alloyed with at least 10–13 percent chromium for corrosion resistance (sources differ on the percentage of chromium required). Nickel may also be added to increase corrosion resistance. Stainless steel jewelry is sometimes identified with the stamp INOX, another adjective used to identify stainless steel; it's from the French *inoxydable* for "rustproof."

"Surgical steel" is a type of stainless steel in varying degrees of hardness that resists corrosion and scratching and is easy to sterilize. It typically contains chromium for corrosion resistance, nickel to provide a smooth and polished finish and molybdenum to make it harder and better able to maintain a cutting edge.

Pat Pruitt, a metalsmith specializing in stainless steel, says that in general, stainless steel can be categorized into two groups, austenitic, and ferritic based on their molecular structure. For jewelry purposes, it is the austenitic group (or 300 series) that is the most common. There is a dizzying array of stainless steel grades out there. Grade 304 and Grade 316 stainless steel are the two most common stainless steels produced. The Grade 316 family is the most common alloy used in jewelry production today and has a couple variants within its alloy family, which Pruitt describes below:

Grade 316 is considered to be the most corrosion resistant of the stainless

Grade 316L is the same as 316 but with a lower carbon content

Grade 316LVM is the same as 316L but smelted in a vacuum environment.

It's not easy to characterize stainless steel because of the wide range of different alloys. In general, most stainless steel jewelry:

♦ resists corrosion and does not discolor the skin

♦ is relatively light weight (specific gravity: 7.7 or 7.8)

♦ wears better than brass and most gold

♦ is a very affordable metal to work with compared to gold and silver

♦ is suited to innovative finishes and creative designs

A down side to working with stainless steel is the toughness of the metal; standard jewelers tools are going to be taxed working with this metal, and that might prove challenging to most artists. Another drawback is that stainless steel alloys containing nickel can cause an allergic reaction in people that are sensitive to nickel.

Fig. 10.7 Grade 316 stainless steel and inlaid fine silver. *Cuff and photo by Pat Pruitt.*

Fig. 10.8 stainless steel bands fused to platinum bars holding tension-set diamonds. *Rings and photo from Stephen Vincent Design.*

Fig. 10.9 Wood grain Damascus stainless steel and 18K gold ring by Chris Ploof Design. *Photo by Robert Diamante.*

Fig. 10.10 316/304 stainless Damascus and 24K gold inlay. *Wedding bands and photo by Pat Pruitt.*

Fig. 10.11 Radial Damascus steel ring inlaid and lined with 18K red gold by Stephen Vincent. *Photo from Stephen Vincent Design.*

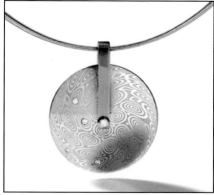

Fig. 10.12 Rosewood Damascus steel and 18K red gold pendant by Stephen Vincent. *Photo from Stephen Vincent Design.*

Fig. 10.13 Damascus steel blade and handle made of antique ivory, 14K, 18K and 22K yellow gold, diamonds, emeralds, sapphires and boulder opal. The sole authorship is by Dellana, meaning she forges the Damascus, fabricates the dagger, sets the stones and does the carving and engraving. Dellana's knives start at $8,000, without the gold and gemstones. The overall length of the dagger is 7 ½". *Photo by Eric Eggly/Point Seven Photography.*

Rising gold and platinum prices have motivated jewelry manufacturers to consider stainless steel as an alternative white jewelry metal. They can buy high quality material and combine it with precious metals and gems to create attractive jewelry. An increasing amount of stainless steel jewelry is now available in stores.

For detailed information on the properties of various steel alloys go to:

www.engineershandbook.com/Tables/steelprop.htm and

www.engineershandbook.com/Tables/hardness.htm

Damascus Steel

Historically, Damascus steel was a type of high carbon steel used in swordmaking from about 300 BC to 1700 AD; it was also known as Wootz steel. Noted for their sharpness and toughness, Damascus steel blades originated in India, where bits of iron and carbon were heated at high temperatures in a crucible and formed lines on the surface of the steel, creating a swirling effect. Afterwards the steel was shipped to Damascus, Syria, where bladesmiths learned to forge it into swords displaying a characteristic surface pattern.

Most of the Damascus steel sold today is an exotic pattern-welded steel, created by welding together from fifty to several thousand layers of different alloy steels and etching it in acid. The acid etches the different alloys at different rates thus creating the wood-grain type pattern. The alternating layers of different steel alloys create a kind of microscopic serrated edge that contributes to Damascus steel blades being superior cutting tools. In addition, the properties of the correct steel alloys make the Damascus steel capable of being hardened and tempered for a very strong, sharp, durable edge and overall blade.

The wood-grain patterns of Damascus steel blades resemble those found on the mokume gane swords of the Japanese samurai. Mokume gane is an ancient Japanese technique of making wood-grain patterns using layers of contrasting colored metals such as copper, silver, gold, palladium and/or platinum.

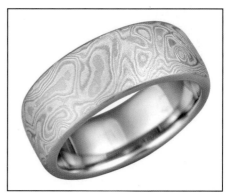

Fig. 10.14 Mokume gane band made of layers of platinum, sterling silver and 14K palladium white gold. *Ring & photo courtesy Krikawa.*

Fig. 10.15 Blade with 324 layer ladder pattern Damascus 203E and W2 steels. Handle: 14K and 22K yellow gold, 14K green gold, 14K rose gold, diamonds, emeralds, black opals, and mother of pearl. *Sole authorship of folding knife by Dellana. Photo credit: Sharp by Coop.*

The mokume gane and Damascus steel techniques are appropriate for wedding rings because they embody the symbolism of marriage. There are two individual elements, distinct looking metals, which are perfectly intertwined. The individual elements follow a path together but never lose their unique identity.

In the 1970's, pattern-welded Damascus steel gained a resurgence in Europe and the USA. It is now available from private bladesmiths as well as companies specializing in creating and/or selling it. The steel is sold by the inch, price depending on steel alloys used, pattern involved and thickness of the metal being purchased. There are also decorative Damascus steels made from alloys that have too low a carbon content to be considered good edge holding steel. Some Damascus steels are made into highly prized jewelry.

Dellana, a specialist in forging and patterning Damascus steel, says that "not all Damascus steels are created equal. It is a process that must be done with skill, as it is very easy to end up with delaminations that reduce the quality both aesthetically and functionally. Some smiths are fine with using metal with flaws in it, but I am not. As in all things, there are very skilled Damascus steel forgers and those that are less so to varying degrees."

Figs. 10.16–10.18 Tungsten carbide rings and images courtesy Poag Jewellers and CrownRing.

Figs. 10.19–10.21 Tungsten carbide rings and images courtesy Lashbrook Designs.

Tungsten and Tungsten Carbide

In the past, tungsten, a light gray metal, was primarily used in jewelry as an alloying element to increase durability. For example, when platinum is alloyed with tungsten, it wears better and is more scratch resistant than other platinum alloys. However, tungsten in the form of tungsten carbide is being used more and more as an alternative material for wedding bands.

In Swedish, tungsten means heavy stone (tung sten). Tungsten was discovered in 1783 by the Spanish chemists Jose and Fausto Elhuyar in the mineral Wolframite. That's why it sometimes referred to as "Wolfram" and has the chemical symbol W. Wolframite and scheelite are two of its important ores.

Tungsten has the highest melting point of all metallic elements and is used to make filaments for incandescent light bulbs, fluorescent light bulbs and television tubes. Tungsten is also used as heating elements in electric furnaces and for parts of spacecraft and missiles which must withstand high temperatures.

All tungsten rings are a combination of elements, usually tungsten and carbon, which forms a compound called **tungsten carbide (WC)**. It is so hard that it's used in place of diamond for cutting tools. The jewelry industry usually refers to tungsten carbide as tungsten and the industrial industry refers to it as carbide. It is made by grinding tungsten and carbon into a powder and then compressing it into a metal mold with high heat and pressure. The compressed material is then fired in an oxygen free furnace at temperatures above 6000 degrees Fahrenheit. This process is called sintering and creates the composite material, tungsten carbide, which is harder than tungsten. A diamond wheel is used to grind designs on a tungsten carbide blank to created different shapes and styles.

Tungsten carbide:
♦ Is very heavy and dense (s. g. 15.63) compared to iron and stainless steel
♦ Doesn't dent easily or get deformed.
♦ Does not discolor skin.
♦ Is hypoallergenic except for some tungsten carbide jewelry, which may contain the element cobalt and causes allergic reactions in some people.

Figs. 10.22–10.24 Cobalt chrome rings and images courtesy Poag Jewellers and CrownRing.

Figs. 10.25–10.27 Cobalt chrome rings and images courtesy Lashbrook Designs.

♦ Is low-priced. Labor costs are normally higher than the metal itself, partly because special skills and equipment are required.
♦ Is more scratch resistant than precious metal alloys.
♦ Can be combined with gold or platinum for interesting designs

Clean tungsten carbide rings with warm water and a mild soap. It is best to avoid ultrasonics and harsh chemicals. Avoid hard knocks because tungsten carbide rings can crack and even shatter if exposed to an extreme blow.

Tungsten carbide rings cannot be resized, but most companies have an exchange policy, which is free or only requires a minimal fee plus shipping if a new size is needed. Nevertheless, if you're looking for a scratch resistant ring similar in weight to gold or platinum, but at a much lower cost, then tungsten jewelry could be an alternative choice for you.

Cobalt-chrome

Wedding bands advertised as cobalt rings are actually made of cobalt-chrome, an alloy made of cobalt (Co), about 27–30% chromium (Cr), molybdenum (Mo) and small amounts of other elements. It's basically the same hypoallergenic metal used for orthopedic implants.

Cobalt-chrome is noted for its superior strength and resistance to scratching compared to precious metals. It's whiter and brighter than tungsten and titanium and resembles platinum, so it is becoming a popular alternative jewelry metal. Be aware that cobalt chrome rings cannot be sized and that a special cutting tool is required if they need to be removed due to hand injury or swelling.

Cobalt-chrome bands come in a wide variety of styles, six of which are shown in figures 10.22–10.27.

11

Niobium, Titanium & Aluminum: Colorful Alternative Metals

Niobium and titanium can be colored brown to blue, yellow, pink, purple, turquoise and then green by gradually increasing the voltage of an electrical current through them while the metals are in contact with a conducting solution. This electrolytic process is called **anodizing**. During the process, a controlled current creates a transparent oxide film on the metal's surface, causing light interference as it is reflected both from the surface of the oxide and the surface of the metal. The appearance of color perceived by the eye is determined by the thickness of the layer and the surface finish. The voltage used determines the thickness of the oxide deposit and the apparent color obtained. Brown colors require the least voltage (about 10-15 volts) and green the most (85–90 volts). Since the colors are created by light interference and not dyes, the colors will resemble those seen in soap bubbles or oil floating on water. Therefore, don't expect to see black, red or dark orange coloring on metal that is only anodized. It is possible, however, to create charcoal-black niobium by heating it red hot.

Other related metals such as tantalum, aluminum, and zirconium can also be colored by anodizing, but titanium and niobium are more accessible and used more often for jewelry. Titanium can also be colored by heating, but it's difficult to control the amount of heat and produce the right thickness of oxide in order to obtain a particular color. It is easier to get the desired color by controlling the voltage during anodizing.

Titanium and niobium are produced commercially, which means initial ingots will weigh in excess of 10,000 pounds. The individual wrought products are made in small quantities of 50 pounds or more and sold by the pound. To simplify the purchase process, the product may be sold by the square inch and foot to individuals.

Niobium

Niobium (Nb) is a lustrous and malleable gray metal that is noted for the beautiful colors it can obtain during anodizing. It was discovered in 1801 by the English chemist Charles Hatchett, who named it columbium. Forty years later the metal was rediscovered by the German chemist Hienrich Rose; he named it niobium, after *Niobe*, the daughter of the Greek mythological Tantalus because niobium is closely related to the metal tantalum. It wasn't until 1950 that the name "Niobium" was officially adopted by the International Union of Pure and Applied Chemistry.

Brazil is the leading producer of niobium, which has a variety of uses beyond jewelry making. It strengthens steel and is important for jet engine components and the nuclear industries. Niobium is added to glass to give it a higher refractive index and allow the optical industry to make thinner corrective lenses. Niobium is also used in pacemakers and for the superconducting magnets in MRI scanners.

Fig. 11.1 Jewelry ensemble made by Brian Eburah with niobium, silver, gold and green tourmalines. *Photo by Peter Mayes.*

Fig. 11.2 Chainmail niobium necklace & photo by Spider of Spiderchain Jewelry.

Fig. 11.3 Brooch by Brian Eburah made with niobium, silver, gold and pink tourmaline. *Photo by Peter Higgs.*

Fig. 11.4 Jewelry ensemble made by Brian Eburah with niobium, silver, gold and blue topaz. *Photo by Peter Mayes.*

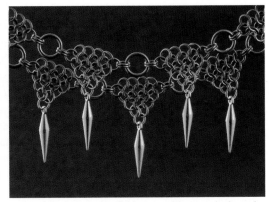

Fig. 11.5 Chainmail niobium necklace and photo by Spider of Spiderchain Jewelry.

Fig. 11.6 Crystalline titanium and silver metal clay pendant and photo by Holly Gage.

Fig. 11.7 Anodized titanium suite by Mr. Titanium. *Photo by Dan Klarmann.*

Fig. 11.8 Titanium band with brushed finish and raw diamonds. *Ring by Todd Reed; photo by Brian Mark.*

Fig. 11.9 Fire colored titanium dial inlaid with 24K gold & hand-engraved with portraits of the jeweler's daughters. *Dial, case & photo by James Roettger in Minneapolis.*

Titanium

People who want sturdy, comfortable jewelry are attracted to titanium. The exceptional strength and light weight that makes it ideal for the aerospace industry also makes it useful for rings and watches, which must be able to withstand a lot of wear. Titanium was first discovered in 1791 but wasn't commercially available until the 1940's, when a refining process for it was invented by William J. Kroll. It's been used for jewelry since the 1970's and is becoming increasingly popular, not only as an anodized metal but as an alternative gray metal as well. Its chemical symbol is "Ti."

Bill Seeley, president of Reactive Metals Studio, Inc., says that there are more than 30 alloys of Titanium, three of which are commonly used on jewelry:

Grade #1 Commercially pure. Ductile, easy to form and work with hand tools.

Grade #2 Commercially purer. Less ductile, harder, easier to machine, most available.

Grade #5, Grade #23 6/4ELI. An alloy of titanium with 6% aluminum and 4% vanadium, hard, tough, strong, available.

Figs. 11.10–11.12 Custom titanium bands and images from Exotica Jewelry.

Common Characteristics of Niobium and Titanium

Titanium and niobium:

♦ Are hypoallergenic

♦ Can be transformed into bright colors using an anodizing process

♦ Are very low-priced compared to palladium, platinum and white gold

♦ Resist tarnish and corrosion; don't discolor the skin

♦ Weigh less than precious metals (SG of niobium: 8.57, SG of titanium: 4.51)

♦ Cannot be annealed in open air; they must be heated in vacuum or inert atmosphere.

♦ Are difficult to size. Niobium and titanium rings cannot be sized by the average jeweler; they must be sent to the manufacturer or sized by a specialist in the metals. If your jeweler or the manufacturer will not size the ring, contact Exotica Jewelry in Clarkdale, Arizona. They have a side business in resizing titanium rings

♦ Are harder than gold, platinum, silver and palladium in their pure form according to hardness tables. They both have a hardness of 6, but hardness values can be deceptive. Sandy Boothe of Exotica Jewelry says "in our experience, niobium is not as hard as gold, platinum or palladium in their pure forms or alloys. Titanium in its pure form is harder than any of these metals except some of the white gold alloys. We don't work in aircraft grade titanium but it is almost certainly harder than any precious metal alloys."

Bill Seely of Reactive Metals agrees that hardness values do not reflect the actual working characteristics of niobium. He says that niobium is very slow to work harden and that "when you take a hammer or chasing tool to niobium, you won't believe its hardness numbers."

Caring for Anodized Niobium and Titanium

Jewelry with anodized niobium and titanium requires special care because the metal oxide film of anodized metal can easily be scratched and scraped off. As a result the color may be removed. Therefore, it is usually best to use anodized niobium and titanium for earrings, necklaces, pendants and brooches rather than for rings which are subject to knocks. There are exceptions, however.

Sandy Boothe of Exotica Jewelry Inc. says that when the coloring is restricted to recessed areas of a titanium ring, it can withstand a lifetime of wear, unlike rings with surface coloring. Niobium is seldom used by itself for rings.

Anodized niobium and titanium can be cleaned with rubbing alcohol, Windex or a non moisturizing mild soap and a soft cloth. Do not use metal polishes or abrasive soaps because they can gradually wear away the surface coloring.

Spider, the founder of Spiderchain Jewelry, says that getting a smudge of skin oil, sweat or perfume on anodized niobium will affect the film on the surface, making the color look drab. However, you can make the piece look like new again with a mild dish soap and some running water or Windex.

Aluminum

Aluminum (Al) is the most abundant metal in the Earth's crust and therefore very affordable, but it is usually found combined with minerals in rocks such as bauxite. Compared to other metals it is very lightweight (specific gravity:2.7) and relatively soft (hardness: 2.75 on the Mohs scale). Its hardness and strength are increased by alloying it with other elements such as magnesium, manganese, copper, silicon and zinc.

Like niobium and titanium, aluminum can be anodized to produce colorful jewelry. However, while researching metals for this book, I noticed that titanium and niobium were the metals most frequently used for anodized jewelry, so I decided to focus on them in this chapter. After seeing some attractive aluminum jewelry by John S. Brana, I decided to add a section on it. I asked him why aluminum was not more widely used. Here is his reply:

Anodizing niobium and titanium is a much easier process than with aluminum. All you need to anodize is increase voltage/time in the electrolytic bath to build up oxide layers to form various colors. Anodizing reactive metals such as niobium and titanium is a high-voltage, low-amperage process performed with nontoxic electrolytes. So in practice, you can achieve a rainbow effect on the same piece of jewelry just by varying the rate at which you pull the item out of the electrolytic bath.

In contrast, anodizing aluminum is more labor and time intensive. Aluminum needs to be degreased, deoxidize, anodized, dyed, and sealed. After degreasing and de-oxidizing, the raw aluminum must have its pores opened, usually through a sulfuric acid bath in order for a thick oxide layer to form, so that the aluminum will accept the dye. Once the dye is impregnated into the honeycomb-shaped pores through a heat bath, the color is sealed by boiling the aluminum in hot water or with a sealer. Although you are left with one even colored object, an endless color palate can be achieved through mixing dyes. However, you cannot produce rainbow effects, as with Niobium.

Most artists I know that use anodized niobium or titanium usually are chain mail artists, and in this instance it would make sense in that aluminum jump rings would be way too soft to hold up to any wear and tear (without work hardening – which doesn't lend itself to chain mail design). It's virtually impossible to solder any of these three metals, so with any application requiring jump rings, you're far better off using niobium or titanium.

Another reason would be design aesthetics. If you are looking for gradient color, this can only be achieved by anodizing niobium or titanium.

As far as cost, you can't beat aluminum. For example, 20 gauge non anodized sheet aluminum runs roughly $.07 a square inch versus $3.60 per square inch for non anodized niobium. Even buying ready-made anodized sheet aluminum is still very cost effective at roughly $.20 a square inch.

The reasons I chose to work with aluminum are mainly raw materials cost, the look of both raw and anodized aluminum, the wide spectrum of colors that can be achieved through anodizing, and its weight. It would be an ideal metal if it could be easily soldered.

Examples of John S. Brana's aluminum jewelry are shown in figures 11.13 -11.16.

Fig. 11.13 Fold formed aluminum anticlastic cuff and photo by John S. Brana.

Fig. 11.14 Model wearing aluminum jewelry. *Jewelry and photo from John S. Brana.*

Fig. 11.15 Anodized aluminum earrings and photo by John S. Brana.

Fig. 11.16 Aluminum cuff & photo by John S. Brana.

Figs. 11.17 & 11.18 Front and back view of an aluminum watch, circa 1925, presented to Charles Holland-Moritz, one of the original founders of the Aluminum Company of America in appreciation for his years of service. *Watch and photo courtesy the Holland Family Trust.*

12

Plating, Coating & Enameling

Coating precious metals with glass, paint, metal and other substances has become increasingly common partly as a result of increasing gold prices. For example, 10K gold jewelry may be plated with 18K or even 24K gold to give it a richer look and make it easier to sell. Silver has become a precious metal replacement for people who are unable to afford gold and is sometimes plated with palladium, platinum rhodium or ruthenium to help prevent the silver from tarnishing and to make it appear more upscale. Silver is also coated with colorless substances to help prevent tarnish. The main disadvantage of coatings and platings is that they can wear away and be scratched; nevertheless, jewelry can be replated or recoated. Coated or plated jewelry is also much more difficult to repair and service, since every application that involves heat would require the replating or recoating of the material that has been burnt away.

In this chapter, I generally use the term **platings** to refer to surface coverings of metal and **coatings** as surface coverings with substances other than metal or glass. However be aware that these terms are sometimes used interchangeably. For example, in their *Precious Metals Book,* the World Jewellery Confederation (CIBJO) states: "A precious metal coating or plating is a layer of precious metal or of precious metal alloy applied to all, or part of a precious metal article, e.g., by chemical, electrochemical, mechanical or physical process."

The ancient Egyptians and Romans used a process called **gilding** or **fire gilding** to decorate jewelry and statues with gold. Mercury was combined with gold powder and the mixture brushed onto an object; the mercury was vaporized with moderate heat leaving a thin layer of gold, or in some cases, silver on the object.

Modern electrochemistry was invented by Italian chemist Luigi V. Brugnatelli in 1805. By the mid 1800's the British and Russians were electroplating printing-press plates with copper using an electric current. The thickness of the copper layer depended on the voltage used and the duration of the current exposure. With the invention of electroplating, fire gilding gradually disappeared because of the toxicity of mercury vapors.

Besides helping prevent tarnish, platings and coatings serve a variety of purposes.

♦ They give base metals such as steel and brass the look of precious metal.

♦ They help prevent allergic reactions caused by base metals such as nickel, but only if the coating is thick enough.

♦ They provide a homogenous metal color and hide soldering lines and marks.

♦ They add color and contrast to jewelry and are an important design element.

♦ They can make white gold look less yellowish and more like platinum. Rhodium is a common plating treatment for white gold. Rhodium's superior hardness also offers some protection from scratching. Black rhodium and ruthenium are used to provide a durable tarnish-resistant gray-black finish or to add contrast to the piece; they can be used as a final finish or as a mask for selective plating.

The way in which metal plating is named can vary depending on the country in which the jewelry is sold. The U.S. Federal Trade Commission's plating regulations can be found at **www.ftc.gov/bcp/guides/jewel-gd.shtm**

The Guide to the Precious Metals Marking Act and Regulations of Canada are at

www.competitionbureau.gc.ca/eic/site/cb-bc.nsf/eng/01234.html

The World Jewellery Confederation (CIBJO) Precious Metals Book Guidelines, which are applied in most European countries are found at

www.cibjo.org: click on "Blue Books" and then "The Precious Metals Book."

Terms & Markings Related to Precious Metal Surface Coverings

As you shop for chains and jewelry, you will encounter a variety of terms indicating the type or thickness of the plating or coating. Here are some of them:

Bonded gold and silver: An unofficial trade term used to describe a layer of at least 10K gold that is mechanically affixed to sterling silver. Example marking: 925 1/20 10K for sterling silver with a 10K gold layer that is 1/20th of the gross weight of the article.

Gold electroplate (GEP), electroplaqué d'or: Having an electroplated gold layer of at least .175 microns or 7/1,000,000 of an inch thick in the U.S. and one micrometer in Canada. The gold must be at least 10K gold.

Gold filled (GF), doublé d'or: Having a solid layer of at least 10K gold bonded with heat and pressure to a base metal such as brass. The Gold layer must be at least 1/20th of the weight of the metal in the entire article. Example markings "1/20 14K GF" or "14 Kt Gold Filled." Gold-filled items, even when worn daily, can last several years, but will eventually wear through.

Fig. 12.1 Mark for 12K gold-filled. *Photo: Renée Newman.*

Gold flashed or **gold washed**: Having a thinner gold layer than 7/1,000,000 of an inch thick

Gold overlay: Having a thinner solid, mechanically bonded layer of gold than gold filled

Gold plate (GP): Having a gold layer of ½ microns or about 20 millionths of an inch. Example markings: "2 microns 12 K. gold plate" or "2μ 12 K. G.P." for an item plated with 2 microns of 12 karat gold.

Rolled gold plate (RGP) plaqué d'or laminé: Having a solid layer of gold less than 1/20 of the gross weight of the entire article. The layer is applied by a mechanical process with at least 10K gold. Marking examples: "1/40th 12 Kt. Rolled Gold Plate" or "1/40 12 Kt. RGP"

Fig. 12.2 Jewelry being electro-plated inside solution at a Rio Grande booth. *Photo © Newman.*

Fig. 12.3 A Frédéric DuClos sterling silver cuff plated with ruthenium. *Image from Frédéric DuClos.*

Fig. 12.4 Sterling silver earrings plated with black rhodium. *Earrings and image from Metal Marketplace International.*

Fig. 12.5 Frédéric DuClos sterling silver earrings plated with rhodium. *Image from Frédéric DuClos.*

Fig. 12.6 Sterling silver earrings plated with 14K yellow and rose gold. *Earrings and image from Metal Marketplace International.*

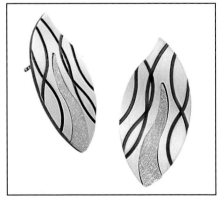

Fig. 12.7 Sterling silver pendant plated with 14K rose gold. *Jewelry & image from Metal Marketplace International.*

Fig. 12.8 Sterling silver earrings by Frédéric DuClos bonded with 18K gold leaf. *Image from Frédéric DuClos.*

Silver electroplate, silver plate, silver plated, electro-plaqué d'argent, plaqué d'argent: In Canada, these quality marks can be applied to an article which has had silver of at least .925 quality applied by a plating process. The thickness of the silver plate must be at least one micrometer or larger. U.S. FTC guidelines state that "it is unfair or deceptive to mark, describe, or otherwise represent all or part of an industry product as plated or coated with silver unless all significant surfaces of the product or part contain a plating or coating of silver that is of substantial thickness." "Substantial thickness" is not defined.

Vermeil: Consists of a base of sterling silver plated with a layer of at least 10K gold and a minimum thickness throughout equivalent to 2 1/2 microns (or approximately 100/1,000,000ths of an inch) of fine gold in the US. The layer of gold may be either electroplated or mechanically bonded. In Canada the thickness of the gold layer must be at least one micrometer.

In general, coatings and platings are supposed to be disclosed. Section 5.16.2.2 of the CIBJO Blue Book states that "when a coating is applied which **changes the colour** of the precious metal alloy used to make the article of jewellery, then the coating **must be declared,** e.g., when using gold coatings on silver; ruthenium coatings on any precious metal . . ."

Section 5.16.2.3 states "When a coating is applied that is the **same colour** as the alloy used to make the article of jewellery, then it is **recommended** that the coating be declared, e.g., when using rhodium coatings on white gold or silver."

E-Coating

E-coating is an electrically-applied paint coating, which was originally developed for the automotive industry. It coats any material that conducts electricity and is being used increasingly to add color to jewelry metals or to provide a tarnish resistant transparent colorless coating to metal such as silver. Coatings can help prevent corrosion; and certain types such as those produced by Aradierung® can also reduce the visibility of fingerprints, make a jewelry piece easier to clean, and/or offer scratch protection by increasing surface hardness.

Pretreatment cleaning is the same as that used for rhodium plating; the piece is polished to remove surface imperfections and cleaned thoroughly. During the e-coating process, a paint or ceramic coating is applied to a jewelry piece at a certain film thickness, regulated by the amount of voltage applied. After the coating reaches the desired film thickness, the coating process slows down. As the piece is taken out of the bath, paint solids cling to the surface and have to be rinsed off. After the post rinses, the piece is placed in a curing unit. The curing unit, a device much like a bake oven, cross links and cures the paint film to assure maximum performance properties. The minimum bake schedule is 20 minutes with the part temperature at 375°F for most e-coat technologies. E-coating equipment and solutions are shown in figure 12.9.

Enameling

Traditional enamel is a form of glass. Metal is enameled by putting a thin layer of enamel powder on it and then firing the piece in a special oven (kiln) until the enamel melts and fuses with the metal. The term "enamel" comes from the High German word *smelzan* (to smelt) via the Old French word *esmail*.

Fig. 12.9 E-coating equipment, solutions and e-coated jewelry at a Rio Grande booth. *Photo © by Renée Newman.*

Fig. 12.11 An artisan kiln from the Evenheat Kiln Company, which can be used for firing precious metal clays, enamels, mokume and low fire ceramics. *Photo from FDJ On Time.*

Fig. 12.10 Enameled frog brooch and photo from Fred Rich Enamel Design.

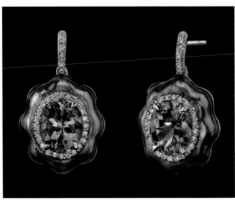

Fig. 12.12 Peridots enhanced with enameling and diamonds. *Earrings & photo from Hubert.*

Fig. 12.13 Blue topaz enhanced with enameling & diamonds. *Earrings & photo from Hubert.*

Fig. 12.14 French enamel bird brooch (circa 1960). *Jewelry and photo from Adin Fine Antique Jewellery.*

Fig. 12.15 Enamel clown brooch (circa 1960). *Jewelry and photo from Lang Antique and Estate Jewelry.*

Fig. 12.16 Enameled portrait on an 18K gold hunter case pocket watch (circa 1890). *Watch & photo from Lang Antique & Estate Jewelry.*

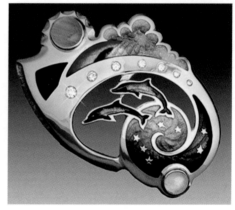

Fig. 12.17 Champleve enameled brooch done by Kristin Anderson of Kristinworks Designer Jewelry. *Photo by Michael Athorn.*

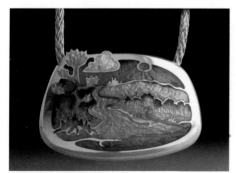

Fig. 12.18 Champleve enameled pendant done by Kristin Anderson of Kristinworks Designer Jewelry. *Photo by Michael Athorn.*

Fig. 12.19 Enameled bracelet. *Jewelry and photo courtesy Denney Jewelers and Belle Etoile Jewelry.*

Fig. 12.20 Plique-à-jour enamel brooch and photo from Lang Antique and Estate Jewelry.

Fig. 12.21 Enamel and aquamarine pearl enhancer & pin by Fred Rich. *Necklace & photo from Fred Rich Enamel Design.*

Metalsmith Kathrin Schoenke says that today's cold enamel materials, such as Novorite, Colorit and many others are resin and hardener-based materials that have the visual properties of glass enamel, but are much easier to apply and repair. Process control is also easier than with conventional enamel.

According to the *Jewelry Making Manual* by Sylvia Wicks, the oldest surviving examples of enamel are gold ornaments with blue glass fused into depressions; they date from 1400 BC. Between 1100 BC and 800 BC, enamel was used mainly to decorate lower quality pieces. Higher quality jewelry was inlaid with stones, usually with garnets. As garnets became more scarce, enamel was often used to fill the cloisons (enclosures) that previously would have held stones, a technique called **cloisonné enameling.** Another technique was **champlevé** ("raised field") **enameling,** in which depressions or cells are carved or cast into the piece and filled with enamel.

Gradually enameling became regarded as a desirable form of decoration in Asia, Europe and the Middle East, even for high quality pieces. During the 1300's in Paris, a new technique evolved from cloisonné enameling called **plique-à-jour** (loosely translated "light of day") where the base plate was left off the piece so that the cloisons filled with enamel resembled stained glass. Enamel portrait jewelry was popular in the 1800's and became a favorite technique of the Art Nouveau jewelers. Recently, it has been revived as a way to highlight gemstones and add color and attractive designs to the surface of jewelry.

Enamel may be transparent, opaque, translucent or opalescent and can be applied to most metals. It's smooth, easy to clean and has a Mohs hardness of 5-6, making it harder and more scratch resistant than silver gold, platinum and palladium. A downside is that it can crack or chip if the mounting is knocked, stressed or bent. Therefore, it is best to wear enameled jewelry as pendants, earrings, brooches and necklaces.

13

Responsible Mining & Manufacturing

This book would be incomplete, if I did not mention steps that the metals industry is taking to conserve natural resources and protect the air, land and water supply. Many companies have come to realize that it is best for them to take the initiative to find ways to avoid polluting the environment and depleting our limited natural resources instead of having the government step in and create laws and bureaucratic paperwork that stifle business.

Many consumers prefer to do business with ethical companies and jewelers who care about the environment and social issues. However, it's not always easy to identify responsible business people. One of the goals of this chapter is to list a few ways that corporate responsibility can be determined. Granted, it is not always possible to prove, for example, that metal is recycled or comes from a country that does not abuse its citizens. Nevertheless, it is possible for a company to establish a paperwork trail, showing what is called "due diligence" in obtaining metals and jewelry from legitimate sources.

Beyond Profits

Responsible companies have higher goals than simply increasing profits. They also aim to help create a better world to live in.

When the mining and jewelry manufacturing industries act responsibly, they can play a major role in improving people's lives. For example, thanks to new finds of gold in Peru, soaring gold prices and responsible business practices, many Peruvians have a higher standard of living, better hospital care, a safer water supply, better homes and more educational opportunities. The production of gold has been the prime reason that the economy of Peru has improved since 2008, in contrast to downturns in the economies of many countries in Europe and states in the U.S.

An example of beneficial and responsible mining in the United States is the Stillwater Mine in southern Montana, about 30 miles north of Yellowstone National Park. It's the only producer of palladium and platinum in the U.S. This underground mine is recognized by regulators and environmental groups for its excellent concurrent reclamation activities, wildlife enhancement projects, community support programs, and responsible environmental management. The Stillwater operation includes an off-site smelter in Columbus, Montana with sophisticated pollution control equipment.

Thomas Jansseune PhD, Managing Director, Umicore Precious Metals Canada Inc., says that his company has identified five aspects of being a sustainable/responsible corporation:

♦ Environmental impact

♦ Origin and Chain of Custody

♦ Health and safety

♦ Social: Community Involvement and Employer/Employee Relationships

♦ Ethics

Ideally, one would like to work with a company that acts responsibly in all these areas. The next sections discuss the characteristics of responsible companies in more detail and some ways you can determine if a company meets proper standards.

Environmental Impact

Companies can help protect the environment and natural resources by:

♦ Using recycled materials when available

♦ Promoting efficient use of resources and energy

♦ Restoring the land properly after mine closure by blending the site into the surrounding landscape. For example, removing mining debris, replacing the top soil and planting native scrubs and trees to create an aesthetically pleasing site.

♦ Using alternatives to hazardous substances in production processes wherever technically and economically viable

♦ Properly disposing of waste. For example, removing all harmful materials from waste water before it is evaporated and refraining from discharging any waste water into the environment

♦ Using pollution reducing technology and equipment. For example, installing recycled, no-loss water cooling systems for smelting furnaces

Two certifications that exist for compliance with environmental protection policies are:

ISO14001: **International Organization for Standardization's** certification for effective and efficient environmental management. This is well established internationally and used in various industries. Companies have to demonstrate that they comply with the law and continuously improve their environmental performance. For more information, go to **www.iso14000-iso14001-environmental-management.com/**.

SCS: **Scientific Certification Systems**. Based in Emeryville, CA, this company identifies itself as a global leader in independent certification and verification of environmental, recycling and food purity claims. SCS certifies producers of metal as a **Certified Responsible Source**. For more information, go to **www.scscertified.com**.

Ethical Metalsmiths, **www.ethicalmetalsmiths.org**, is another organization that promotes environmental and social responsibility. Numerous countries and various organizations give awards for environmentally friendly companies.

Origin and Chain of Custody

Ethical companies and precious metals organizations want to disassociate themselves from money laundering and the trade in conflict minerals. For example, in 2012, the World Gold Council established official guidelines for a **Conflict-Free Gold Standard**. The objective of their Standard is to create absolute trust that the gold produced under its guidelines neither fuels armed conflict, nor funds armed groups, nor contributes to human rights abuses associated with these conflicts. The Standard is designed to apply globally for World Gold Council members and other companies involved in the extraction of gold. You can get details of their Conflict-Free Gold Standard at **www.gold.org/about_gold/sustainability/conflict_free_standard/**.

In March 2012, the Responsible Jewellery Council launched its **RJC Chain-of-Custody (CoC) Standard** and supporting documents for the precious metals supply chain. Its purpose is to help identify and promote gold and platinum group metals that are conflict free and responsibly produced and traded through the jewelry supply chain. It goes beyond environmental issues. You can get more information about the RJC chain-of-custody certification program at **www.responsiblejewellery.com/chain-of-custody-certification/**.

The Scientific Certification Systems SCS offers a Recycled Content Certification, which helps prove that products are recycled.

Health and Safety

Responsible companies provide a safe work environment for workers at mines and factories. The Occupational Health and Safety Zone OHSAS is an international health and safety management system that assesses and certifies companies under OHSAS 18001 so as to help minimize risks to employees/etc. For more information, go to **www.ohsas-18001-occupational-health-and-safety.com/what.htm**.

Social Responsibility

Being socially responsible involves:

♦ Paying workers a living wage and treating them fairly and with respect

♦ Supporting the development of mining and manufacturing communities by building roads, hospitals and schools and whatever else is necessary to help reduce poverty

In numerous countries, various organizations give "Preferred Employer" awards which also cover social responsibility to some extent. The ISO has launched the development of the 26000 standard providing voluntary guidance on social responsibility. See **www.iso.org/iso/socialresponsibility.pdf**.

Fairtrade International aims to ensure that producers receive prices that cover their average costs of sustainable production and that the conditions of production and trade of all Fairtrade certified products are socially, economically fair and environmentally responsible. See **www.fairtrade.net/aims_of_fairtrade_standards.html**.

The Fairtrade certification system is run by a separate company called FLO-CERT. See **www.fairtrade.net/certifying_producers.html**.

Ethics

Responsible companies treat their customers fairly, label their products accurately and provide proper disclosures. In other words, they follow the golden rule of "Do unto other as you would have them do unto you."

Some organizations involved in setting standards for selling precious metals are:

♦ The World Jewellery Confederation (CIBJO). To get a link to their *Precious Metals Blue Book*, go to **www.cibjo.org**.

♦ Jewelers Ethics Association, **www.jewelersethicsassociation.com**

♦ Jewelers Vigilance Committee in the U.S. (JVC), **www.jvclegal.org**

♦ Jewellers Vigilance Canada, **www.jewellersvigilance.ca**

♦ Convention on the Control and Marking of Articles of Precious Metals, also known as the Hallmarking Convention, **www.hallmarkingconvention.org**

♦ The London Bullion Market Association, **www.lbma.org.uk**

♦ International Organization for Standardization (see ISO 9001) **www.iso.org**

Small companies normally cannot afford the expenses involved in third party certification and audits. There are still ways of determining their commitment to being a positive force in society: Examine their websites and mission statement; get references; contact the better business bureau; visit the company in person; talk to their employees; examine the forthrightness of the terminology they use to describe their products; and ask the owners and/or managers about their recycling policies, the origin of their products, their community involvement and their goals. Many have received community award certificates and they will gladly show them to you.

14

How Much is Your Jewelry Worth?

The least expensive way to find out the liquidation value of the metal in your jewelry, is to ask some gold buyers or jewelers how much they'll give you for it. If it's an estate or jewelry piece, you should also contact antique specialists and/or auction houses to find out if it has value as a collectible or antique piece. If the piece contains expensive gems, it's best to consult a knowledgeable jeweler, gemologist or auction house because most gold buyers and pawn shops will pay little or nothing for the gems. In many cases, they will even charge you for extracting the gems if the gold is melted.

Nowadays, it's easy to find buyers for gold mountings and chains. Jewelers, pawnbrokers and coin shops typically buy gold as well as other precious metals. In addition, you can do an Internet search by typing in "gold buyer" and your city. While driving, you are likely to notice signs that say "We buy gold," especially in jewelry and pawn shop districts. This chapter focuses on gold rather than platinum, palladium and silver jewelry because gold jewelry accounts for most of the second-hand jewelry market. Keep in mind that the principles for buying and selling silver, platinum or palladium are the same as those for buying gold.

Don't expect to get what you paid for gold jewelry unless there has been a major increase in the value of gold since it was purchased. Taxes and manufacturing costs cannot be recovered. For example, when I was in Germany in 2012, I asked a Dubai seller of 22K gold jewelry about the difference in his selling and buying prices. He said:

> Our 22K gold jewelry is selling for 58 Euros ($75) per gram today. However, if you wanted to sell it back to us today, we would only buy it back for 36 Euros ($46) per gram because we wouldn't pay you for the fabrication costs, the value-added tax of 19%, or the cost of the metals with which the gold was alloyed. In addition, the buying price of gold is a little lower than the selling price.

Before you show your gold jewelry to buyers, it's a good idea to calculate its **scrap value** (the value of the materials it's composed of). In the case of gold jewelry, alloying metals such as copper are not included in the scrap value. Only the value of the gold is calculated. To determine the scrap value, you'll need to weigh your jewelry. If a gold scale or balance is unavailable, try using a postal scale. Despite their limited precision, postal scales can give you a rough estimate of gold weight in grams or in ounces avoirdupois (ounces avoirdupois are used when weighing food, people, letters, etc. Gold weight is measured in ounces troy, which are about 10% heavier).

Real examples are more fun to work with than theoretical ones. The gold-buying example I've chosen for this section is the same as the one I used in my *Gold Jewelry Buying Guide*, which was published in 1993. The principles for determining the scrap value of gold jewelry today are the same as they were in the 1990's, however gold prices today are several times higher, and more people are offering to buy your gold.

The hollow San Marco bracelet in figure 14.1 broke two weeks after it was purchased for $120 in 1993. The owner, Mrs. Smith, was told by a repair jeweler that the bracelet was not worth repairing because it could break again after it was fixed (the store did not offer refunds or exchanges for broken jewelry). Mrs. Smith asked me to help her sell it and I gladly did so because I thought it would be a good example for a chapter on selling your gold jewelry. I took the bracelet to some dealers in downtown Los Angeles to see what they would offer for it. The results are discussed in this chapter.

Fig. 14.1 San Marco chain. *Photo © R. Newman.*

The weight of Mrs. Smith's bracelet is **8.2 grams** (5.3 pennyweights). It is **14K**, which means that it's supposed to contain 58.3% (.583) gold. To determine the cash value of the gold in the bracelet, we need to know the current price of gold at which the major gold exchanges are listing and selling their gold; it's quoted in terms of the troy ounce.

On the day I showed Mrs. Smith's bracelet to dealers, the current price of gold was **$327.50/oz t**. (The Internet or the financial pages of newspapers or brokerage firms can give you the current prices of gold. In this case, I checked a display board at a gold exchange in Los Angeles). Many precious metals sites also offer gold price calculators. However, it's helpful to have an understanding of how gold values are determined.

The formula below will give us the value of the gold in a 14K gold alloy.

Spot gold x gold content (.583) = gold value

$327.50/oz t x .583 = **$190.93/oz t**, the value of one troy ounce of 14K gold that day

Since the weight of Mrs. Smith's bracelet is given in grams (or sometimes pennyweight in the U.S.) instead of troy ounces, we need to know how many grams are in one ounce troy. This and other weight equivalencies are given below (**g** = grams, **oz t** = ounce troy, **oz av** = ounce

Table 14.1 Weight Conversion Table

Unit of weight	Converted weight
1 pennyweight (dwt)	= 1.555 g = 0.05 oz t = 0.055 oz av = 7.776 cts
1 troy ounce (oz t)	= 31.103 g = 1.097 oz av = 20 dwt = 155.51 cts
1 ounce avoirdupois (oz av)	= 28.3495 g = 0.911 oz t = 18.229 dwt = 141.75 cts
1 carat (ct) (gemstone weight)	= 0.2 g = 0.006 oz t = 0.007 oz av = 0.1286 dwt
1 gram (g)	= 5 cts = 0.032 oz t = 0.035 oz av = 0.643 dwt

Knowing that 1 gram is equivalent to 0.032 oz t, we can calculate the gram price of the gold in the bracelet as follows:

0.032 x $190.93 = **$6.11/g**

Mrs. Smith's bracelet, which is entirely gold and has no gemstones, weighs 8.2 grams so the gold value of her bracelet is $50.10.

gram weight x gram price = value of gold in bracelet

8.2 g x $6.11 = **$50.10**.

If you don't have a calculator, it will be easier to use rounded figures such as $330 for the current gold price. You can round .583 down to .58 or lower.

The results of an assay (test that determines the gold content) of one ounce of 14K gold normally will not be .583. In countries such as the United States where the use of low karat solder is permitted and where enforcement of the plumb gold law is minimal, the gold may also be underkarated, especially if it is an older piece. According to Thomas Jansseune PhD, Managing Director, Umicore Precious Metals Canada, the standards in the USA changed in 1976; before that date there was a higher tolerance on the gold content than there is now. This means that older pieces may contain less gold than newer pieces even if they are both stamped 14K (½ karat for pieces without solder, 1 karat for pieces with solder). Standards in Canada changed in 1981. In addition, there may be repairs with other material or solder and even fake material.

If we round the gold content down to 55% and calculate the gold value of the bracelet on the basis of a 55% gold content, it would be about $47. Nobody buying it for scrap, though, would pay that amount for it. There are several reasons for this: A profit margin has to be figured into the offer. It costs money to have scrap **assayed** (analyzed to determine its components) and **refined** (separated into its component metals). There are costs involved with respect to trading the gold and any exchange rate differences. The piece may contain less gold than expected. If the item turns out to be stolen, it could be seized by the police. In certain areas, dealers must wait a month before they can melt or sell secondhand merchandise. This gives the police time to determine if it's been stolen. Meanwhile, the buyer's money is tied up. The higher the buyer's risks, the lower his offer will be.

Mrs. Smith's bracelet was taken to six places that deal in gold. The first was a pawn shop. They said they only loaned money and sold gold jewelry. It's unusual for a pawn shop that deals in gold jewelry to refuse to buy it. One other place said they wouldn't buy the bracelet—a gold exchange which did not deal in jewelry. They recommended an assayer and said they would pay 2% below the current price of ($327.50) for any gold accompanied by an assay certificate. I doubt these refusals to buy gold would have occurred in 2012. Times have changed and gold prices are several times higher.

The offers of the other four places are listed below. Besides buying gold secondhand, all four places sold gold jewelry.

1. $40   2. $40   3. $30

4. $40, first offer. $50, second offer after being told $40 was not enough.

It's surprising that three of the four places made the same offer and that they offered 80% of the current gold price. An offer of about 70% or more would be more common in a low-risk purchase situation. (It was unlikely this bracelet was stolen or underkarated). The higher than expected percentage was perhaps due to high volume and easy access to gold assayers and gold buyers in downtown Los Angeles. When selling your gold, keep in mind that the price paid for scrap gold varies considerably depending on the geographic location and market conditions of your area. Naturally, too, it can vary from one gold dealer to another.

All four places weighed the bracelet and checked its karat stamp and trademark. The last buyer was the only one to do an acid test on the bracelet, measure its length, and examine the broken area. He obviously planned to repair it and resell it as a bracelet rather than as scrap. A retail customer might consider a price around $90 to be a bargain. If the bracelet were designed to last, it would be. The $50 is only the value of the gold. It does not include labor, sales taxes, retail profit, import duties (the bracelet was made in Italy), or the cost of the alloying metals of silver and copper. The original purchase price of the machine-made bracelet from a high-volume discount store was $120 plus sales tax. It's surprising that it could be resold for $50, 41% of its retail price because that would hardly cover the costs.

Some people become disillusioned with jewelry when they discover how little they can resell it for. This is because they've bought jewelry for the wrong reasons or because a salesperson has misled them about its value. If a store tells you their jewelry is worth double the price they're selling it for, don't believe them. They'd be selling it at the higher price if they could find a willing buyer. Think twice before buying anything from them because if they're willing to misrepresent the value of their jewelry, there's a chance they'll misrepresent the quality of their gems and gold.

If investment or resale potential is your only reason for purchasing gold jewelry, it is better to buy gold bars or coins instead. That way you won't have to pay for labor and retail markups or in most cases sales taxes. Nevertheless, in Asian countries, jewelry is regarded as an investment product to be sold in difficult times. Therefore, they have a preference for 22K or 24K gold jewelry.

When jewelry is bought as a clothing accessory, an ornament, or a memorable gift, it can be a good buy. A solid, well-made piece of jewelry will outlast other accessories. Even today, the gold jewelry and artifacts of the ancient Egyptians look as good as new. When people are asked what gifts they treasure the most, more often than not they'll say jewelry. It's no coincidence that the first official exchange of gifts between a husband and wife are their wedding bands. The durability and never-ending circular form of the rings make them the perfect symbol of an eternal commitment of love.

Even when a poor jewelry choice is made, it may turn out to be less of a loss than if it were another type of merchandise. Suppose you bought a $120 figurine, and a few weeks later, you wanted to resell it. If the figurine were broken, you probably wouldn't be able to find any buyers for it. In less than 1/2 hour, I found three shops that were willing to pay $40 cash on the spot for Mrs. Smith's broken bracelet. The bracelet could have also been pawned. The pawn value tends to be about 50 to 75% of what the item can be sold for (except in some states where the interest rate allowed decreases as the amount of the loan increases.)

The way in which this chapter calculated the value of Mrs. Smith's bracelet may lead some people to believe that jewelry value equals gold value. Such a conclusion would be wrong. When assessing jewelry, you also have to consider the craftsmanship, artistic merit, antique value and overall desirability of the piece, the value of diamonds or other precious stones, whether or not the piece is in fashion or trendy and whether or not the piece is collectable. Thus, there is no simple answer to the question, "What is your jewelry worth?" The value of a piece depends on the purpose for which it is appraised (insurance, estate, liquidation etc.) and the subjective opinion of the owner or buyer.

If you have an antique or estate piece of jewelry, its antique value could be worth much more than its intrinsic metal value. Before selling it for scrap, have it checked by one or more antique specialists. They may offer to buy it from you for a much higher price than a gold buyer. Many valuable antique pieces have been melted down, much to the dismay of historians.

Jewelry Rescues:

Spotlight on valuing antique & vintage metalwork

By Sheila Smithie, Antique & Estate Jewelry Specialist and Fellow of the Gemmological Association of Great Britain

Valuable antique and estate pieces have been lost to the scrapper because neither seller, nor buyer, realized that the item had value beyond its metal weight. Before selling heirlooms and antique jewelry for their scrap value, consider having the jewelry evaluated by specialists at an auction house, either by making an appointment to bring in pieces, or sending detailed images. Auction houses normally do not charge for the initial review and verbal appraisal of jewelry, as it is an integral part of their business model. Owners can also send jewelry to reputable major auction houses for review through insured mail if an "in person" visit is not possible.

The heads of jewelry departments of major auction houses, and their teams of specialists, combine to offer many years of experience in their field. Through constant interaction with estate lawyers, collectors and the general public, jewelry specialists review and auction large volumes of old jewelry. They are therefore in an excellent position to recognize jewelry that is collectible but not obviously valuable, even when the items are not signed or even marked for gold content. Specialists can determine jewelry's metal content with non destructive tests, and, as part of their evaluation, can offer a realistic idea of a likely market value, using their knowledge of comparable items that have been recently auctioned.

Consider the following cases of unmarked antique and vintage metalwork, handled by Gloria Lieberman and her department at Skinner Auctioneers:

The antique chain in fig 14.2 is made of a form of brass called "pinchbeck," an alloy of copper and zinc, popular in the 18th century as a substitute for gold. Although the metal content value is negligible, the chain was bought by specialized collectors for $1100.

Fig. 14.2 An antique brass "pinchbeck" long chain bought for $1100. *Photo courtesy of Skinner Inc.*

This modern looking but antique 18K gold (un-marked) wirework choker was found by an estate liquidator under a bed. It sold for $1900 at auction on December 8, 2009. This was more than three times its scrap value for 24.2 pennyweights.

Fig. 14.3 An antique unmarked 18K gold wirework choker sold for $1900 in 2009. *Photo provided courtesy of Skinner Inc.*

The Archaeological Revival 18K gold and amethyst fringe necklace in 14.4 was found tangled up with other items in a plastic bag stuffed with costume jewelry which Skinner reviewed in the thorough process of conducting an estate appraisal. In such appraisals, no item is overlooked. The owners and the heirs were completely unaware of the value of this necklace, which had been handed down through the generations and was somehow forgotten. Its gold content was unmarked, and the double "C" Castellani hallmark was small and difficult to find without an expert to recognize the work and know where to find the maker's mark. Castellani were

Fig 14.4 An antique 18K gold & amethyst necklace that brought $26,000. *Photo courtesy Skinner Inc.*

among the most important jewelers of the nineteenth century, and their jewelry, collected by Americans on Grand Tour, is highly prized today. The necklace, even though damaged by harsh treatment, brought $26,000.

In other cases, workmanship that appears amateur can lead sellers to undervalue items. They can mistake handcrafting for crude construction and poor taste, as with the unsigned Arts & Crafts period bracelet below. Before scrapping the unwanted bracelet, however, the owners had it evaluated among a group of other gold and costume items. Skinner sent the bracelet to Edward Oakes' family for authentication as a 1920s work from their studios, and it brought $5500 at auction.

Fig. 14.5 Arts & Crafts period bracelet by Edward Oakes Studios, c. 1920's, sold at auction for $5500. *Photo courtesy of Skinner Inc.*

Fig. 14.6 An antique Boucheron arm band that brought $34,000. *Photo courtesy Skinner Inc.*

An extraordinary example of fine gold work, whose style fell out of favor in the 20th century, is exemplified by the armband in figure 14.6. It came to Skinner from a pawnbroker and was entirely without hallmarks or signature. Further, it seemed to have been damaged. It had drilled holes that did not serve any purpose, and was missing several small gems. It appeared too large to wear. However, research subsequently showed that the piece was an 18K gold armband from the Parisian jeweler Boucheron, and bought by the American millionaire's wife Marie Louise Mackay in 1876. The drill holes had once accommodated a large detachable brooch, since lost, likely destroyed for the natural pearls with which it was set. Even in this incomplete condition, the armband brought $34,000, almost ten times its substantial weight in gold.

Finally, the large silver brooch in figure 14.7 was brought to the attention of Mrs. Lieberman, whose reputation as an early supporter of artist jewelry during her 40 years in the business is widely known. The owner had purchased the unmarked, unsigned piece at a large garage sale in suburban Los Angeles. Initially reluctant to spend the money on herself, she bought it because it "spoke to her" and had remained unsold after the third day despite the $5 price tag. It was made of a single piece of wire, including the brooch pin. Though the owner did not think to connect the piece to a major modern artist, she nonetheless felt it had intrinsic artistic value and that Mrs.

Fig. 14.7 Calder silver brooch, c. 1942, that sold for $33,000 at auction. *Photo: Skinner Inc.*

Lieberman, whom she knew from Antiques Road Show appearances, would appreciate it. Skinner subsequently handled the brooch's authentication by the Calder Foundation, whose experts dated it to 1942. The brooch brought $33,000 at auction, well beyond its value as a few ounces of silver.

The prior examples, illustrate that the auction market is an important option for sellers to consider. It is vital to understand that presenting jewelry at auction is different from selling jewelry directly to dealers. Since auction houses do not buy jewelry but act instead as the seller's agent, the auction house's interests and the seller's interest are mainly aligned. The auctioneer is paid on commission, which is maximized by achieving the highest possible price for the seller

At the same time, the auctioneer's business depends on describing the items accurately to potential buyers, which helps it achieve a good reputation among buyers and high prices for sellers. This positive dynamic benefits all the participants in the market. Sellers should understand that auction houses charge both a standard seller's commission and an auction premium over the hammer price (paid by the buyer), which are used to cover the substantial marketing and staffing expenses of running an auction. The commission and premium are of course the main source of the firm's revenue. When sellers are auctioning valuable items of jewelry, the seller's commission may be negotiable, but this is not always the case if the jewelry's value is modest. There is often a four to six month time frame before the jewelry can be auctioned and payment made. If the seller needs immediate funds, auctioneers may advance money, but only if important and valuable items are being sold. Each item sold is given an auction estimate based on hammer prices of similar items in past auctions. Specialists should be willing to explain the rationale behind these estimates. Typically, a mutually agreed reserve will be placed on the jewelry to ensure that it does not sell below intrinsic value.

Increasingly, antique and vintage jewelry is gaining appreciation as an important part of our cultural and artistic heritage. New jewelry curatorships have been established at major institutions, such as the Museum of Fine Arts Boston, who are pioneers in this area. Many auction house specialists and dealers are highly motivated to share their knowledge and experience in this cause, and owners can both benefit from and participate in this trend.

Sheila Smithie worked for ten years as a specialist for Gloria Lieberman in the Jewelry Department of Skinner Inc. Auctioneers.

Tips on Reselling Your Jewelry

♦ Have copies available of your purchase receipts, appraisals, credit card purchase statements and/or lab reports. These help verify the quality and identity of the gems and metals, and they help prove you are the rightful owner. The more paperwork you can provide, the lower the buyers' risk, and the more they'll be willing to offer.

Anti-money laundering legislation is in place in various countries, and this may require gold buyers to ask for an ID. Sellers should cooperate and question buyers who don't do this.

♦ Try taking the jewelry back to the store where you bought it. Ask if they will buy it back or offer you credit towards other jewelry. Sometimes the original seller will give you the best offer.

♦ Consider having a jewelry store sell your piece on consignment. Consignment sales sometimes bring the highest price. Always get a written contract and insurance on the item through the store's insurance. Make sure, too, that they put it on display. Occasionally, a store will put consignment pieces in the safe and forget about them.

♦ Get offers from people who understand the value of your jewelry. An antique specialist, for example, will probably have the greatest appreciation for the antique value of a piece. A coin dealer will be the most likely to know the value of a coin pendant. Colored stone experts will offer more for a valuable ruby or emerald than a pawnbroker who doesn't understand how they are valued. In fact, there are many pawnbrokers who won't give you anything for colored stones. They'll only pay for gold, platinum and diamonds. There are others, however, who are very knowledge-able about colored gems and who will offer you a fair price for them.

♦ Get offers from people who sell jewelry like yours at the retail level. The best offer for Mrs. Smith's bracelet was from a retail jeweler who sold and repaired gold chains. The retail jewelry value is higher than the wholesale or scrap value of a piece.

♦ Do not invite strangers to your house to look at jewelry. This could result in a hold-up, burglary, rape or death. If you need to show your jewelry to a buyer you don't know, have him or her meet you inside a bank where you can take it out of a safe deposit box. Even this procedure can be risky.

♦ Do not appear to be desperate for money, even if you are. Usually the more eager you are to sell, the less the buyer will offer.

Characteristics of Ethical Dealers & Consignment Contracts

Before consigning valuable antiques and jewelry, it's helpful to know what you can expect of an ethical consigner and consignment contract. Steve Pierce, a banker who has overseen the management, distribution, liquidation, and operations of several large prestige estates, offers the following information and advice to people who are liquidating their jewelry and estates.

♦ Any antique dealer who advertises for or normally takes consignment goods into his inventory should always have on hand a written contract, which specifies the rights, duties and obligations of both the dealer and the owner to the goods.

♦ The dealer should not only be willing, but also encourage the owner to take a copy of this contract for examination prior to giving possession of the items to the dealer.

♦ If the property owner appears confused or not comprehending, the dealer (for his own protection) should ask that person to bring a friend, relative, or care giver to witness the execution of the contract.

♦ A contract of consignment should contain several essential elements such as the date it is signed and when it will expire or terminate. This contract cannot run in perpetuity—it must have an ending date.

♦ Both a reserve (minimum) price and a list price of the jewelry should be stated. This is the amount that the public will be asked to pay for the item(s).

♦ If the dealer has the option to purchase the items himself, it should be so stated and there should be a time limitation and an agreed amount. That amount could be the reserve or percentage of the list price.

♦ The dealer must state in writing that he will advertise or promote the merchandise in quest of a fair market sale.

♦ The merchandise must be clearly and concisely described, along with the condition of the items. The owner should certify that he or she in fact has clear and unencumbered title to the items and has the legal right to sell them.

♦ The contract should specify exactly who is responsible for transporting the merchandise to and/or from the dealer.

♦ The exact percentage the dealer may retain as his commission from the sale must be spelled out. Normally the dealer retains from twenty-five to thirty-five percent of the selling price. The most commonly accepted commission is thirty-percent.

♦ Normal reporting dates should be established. Usually reputable dealers will report the status of the consignment every thirty days, and pay any amount due at the end of each calendar month.

Selecting a Refiner

If you are a jeweler, hobbyist or even an individual with a lot of jewelry or scrap metal that you would like to exchange for cash, it is to your advantage to sell it directly to a **precious metal refiner**, a facility that separates and purifies metals. Most refiners only deal with members of the jewelry or precious metals trade, but depending on local city/county/state regulations, some will deal with individuals who have a significant quantity of precious metal jewelry to sell. A few refiners have no minimum, but for small amounts of precious metals, it may not be cost effective to deal with a refiner. Dan Kapler, refining sales manager for David H. Fell and Company Inc., suggests selecting a refiner that is willing to work with lower volume customers in an effort to ensure that processing costs will not exceed their expected return.

Refiners differ in their payment policies. Don't assume that the refiners who offer the highest percentage for your gold are necessarily the best choices. It is more important to consider a company's reputation for accuracy than its posted rate of return—numbers are just numbers if not backed up by proper science in the form of detailed, accurate assays. Also look for hidden fees; Kapler advises customers to ask for documentation of all potential costs and payout rates.

If you are interested in finding a large refiner with a wide array of services, consult the International Precious Metals Institute, **www.ipmi.org**. Here are some considerations and questions to ask when selecting a refiner:

♦ **What is their minimum**?

♦ **What percentage of the value of the precious metals do they pay**? As mentioned previously, the highest price is not necessarily the best net value. To make an apples-to-apples comparison, it is important to consider any hidden fees, which market price/exchange rate you will be offered, and the possibility of metal for metal discounts with your settlement

♦ **Can the metal's price be "hedged" at a date of your choosing?** – i.e., locked in on the day of shipping or receiving?

♦ **What types of metals do they process**? Some refiners only pay for gold and perhaps silver. If you have palladium and platinum, consider working with a full service refiner that can process and pay for all your precious metals.

♦ **Do they charge for stone removal**? If so, how much? Do they offer gemstone soaks? Will they return the stones to you or offer to sell them on your behalf? Policies vary from one refiner to another.

♦ **Do they offer to pay for all the precious metals present in your lot**? In the case of palladium white gold, for example, some refiners will pay you both for the palladium and the gold. Others may pay you only for the gold.

♦ **Do they have any third party certification?** One such certification is ISO 9001, which is an audit based on an extensive sample of the company's facilities, products and services.

Preparing Jewelry and Metal for a Refiner

Before sending jewelry to a refiner, read Chapter 8, "Real or Fake" and Chapter 9, "Determining Karat Value and Fineness." You'll need a basic understanding of how to identify precious metals in order to prepare and sort it for the refiner. Here are some recommended steps to preparing your jewelry and other metal items to send to a refiner:

♦ Use a strong magnet to sort out all items containing iron or steel as they will interfere with your assay and settlement. Remove all items that stick to the magnet.

♦ Check the karat and fineness stamps. Sort the pieces according to their markings. Verify if the markings are accurate with the touchstone test and fresh acids, gold testers and/or XRF machines.

♦ Weigh your goods. If you don't have a calibrated scale that is legal for trade, then your figure will only be indicative and the refiner should alert you to any discrepancy upon receipt. Afterwards you should calculate the fine gold content. (Legitimate refiners prefer their customers be knowledgeable and aware of their expected returns.) Remember that there is some weight loss during the melt but that no gold should be lost. (Small amounts of gold can be caught up in the slag produced when adding borax to remove impurities.) At the same time, expectations should be realistic, especially when the weight includes pieces with stones or of unknown karatage or that have been repaired or altered.

♦ Enclose groups of items with the same metal type and fineness in separate double plastic bags and tie and seal them securely.

♦ Find out the current precious metals prices. They're available online.

♦ Include a detailed list of the contents and their weights inside the shipping box with your name, address and contact information and the estimated value of each bag. (Ask the refiner if they have their own preferred packing list.) Retain a copy of the content list.

♦ Photograph the items you are sending and the outside of the box after it is prepared for shipment.

♦ Use bubble wrap, crumbled paper or Styrofoam packing peanuts to prevent the contents from moving about and creating noise inside the box.

♦ Do not use terms such as "gold," "jewelry" or "refinery" in the addresses on the box when mailing precious metals or jewelry. If necessary, abbreviate the name of the company to which the package is being shipped.

♦ Tape and seal all edges of the box.

♦ Insure the package(s) in case of loss. The insurance company will typically provide shipping and packaging requirements such as double boxing, the choice of carrier etc. Registered mail is often the most secure and cost-effective way for small shipments. Read the fine print – even major shipping companies that offer insurance will pay only a set limit for losses involving precious metals, regardless of the premiums you've paid. Ask the refiner if they can provide insurance for your package under their own policy.

♦ Keep both written and digital records of the payout(s) for your shipment(s). The kind of records that one needs to keep will depend on each country's legislation and will often be needed for tax & anti-money laundering (AML) purposes. Legitimate refiners will require strict compliance with AML laws and necessary forms from their customers.

♦ If you're repaid with recycled precious metal and you are a jeweler or manufacturer claiming to sell eco-friendly jewelry, keep records of the source(s) of the metal that you buy to help prove your claims. Refiners can typically return the material not only as product for jewelry applications but also as investment products.

15

Caring for Your Metal Jewelry

Which of the following scratches most easily?

♦ Glass

♦ Gold

♦ Platinum

♦ Turquoise

Pure gold scratches most easily because it's the softest. Glass and turquoise are harder than gold. They're even harder than platinum. This means that if you place jewelry in a box on top of other pieces, it could get scratched, especially by jewelry set with gemstones. Therefore, when you store jewelry, place each piece in a separate compartment, pouch or plastic bag, or wrap them individually in soft material. Padded jewelry bags with lots of pockets are also handy for storing jewelry.

The disadvantage of placing jewelry in cloth pouches or bags is that you can't see it and find it readily. One solution is to use stackable trays with or without compartments or large chests with trays or drawers. Jewelry cases, folding boxes, and rolls that open up with a large display area are also available. You can find these at jewelry supply stores for reasonable prices. It costs even less to place your jewelry separately in small clear plastic pouches and organize them in ordinary boxes. A word of caution about plastic—it may cause silver to tarnish more quickly.

Conventional jewelry boxes can protect pieces from damage if they're stored individually, but these types of boxes are one of the first places burglars look when they break into a house. Therefore, it's best to reserve jewelry boxes for costume jewelry when they're displayed on tables or dressers. Use your imagination to find a secure place in your house to hide jewelry pouches, rolls, cases and boxes. If expensive pieces are seldom worn, it may be wise to keep them in a safe deposit box. Special storage boxes for watches called winders are popular with people who collect watches. These have computer-controlled rotators, which keep the watches in good working condition. They are great gifts for watch connoisseurs.

Cleaning Metal Mountings

The safest way to clean jewelry mountings is to rub them with a soft cloth dipped in warm water containing a mild liquid detergent. Brushes and scouring pads can scratch metals such as gold, platinum and silver. Therefore, avoid using them to clean these metals, especially if the jewelry is plated. Powder cleansers and toothpaste can also wear away metal, so don't use them for cleaning jewelry.

Some jewelers like to clean jewelry with sudsy ammonia or window cleaner sprays. They're usually safe to *spray* on metal and most stones, but they may damage gems such as pearls, coral and turquoise. However, avoid soaking jewelry in ammonia. Overexposure to it can sometimes change or darken the color of certain solders or gemstones.

In the diamond trade, ethyl alcohol is frequently used to clean diamonds because it can dissolve grease build-up, and it evaporates quickly without leaving water spots. Avoid common isopropyl rubbing alcohol; it contains oils that can leave a film residue on your jewelry. Ethyl alcohol can be found in hardware stores, and it should only be used on stones that are not damaged by chemicals.

Removing tarnish, especially on silver, can be a challenge. One way of removing tarnish is to rub the silver with a soft, moist cloth and baking soda, which has a Mohs hardness value of 2.5. This value is about the same or less than that of silver alloys. Consequently, the soda is not as apt to wear away the silver as many other abrasives. In fact, baking soda is safe enough that many dentists recommend it as an ideal cleaning agent for teeth. Rinse the piece well with water so the baking soda does not remain in crevices.

Toothpaste and silver polish pastes can also remove tarnish, but the particles in them are usually harder than those of baking soda, making them more abrasive than baking soda. Silver polishes, however, may contain additional chemicals that make them more effective.

Liquid silver cleaners are also available, but they should not be used on jewelry set with stones such as pearls, coral, turquoise, malachite, etc. Read the labels on jewelry cleaners and polishes. Chapter 5 has other tips on cleaning silver.

Gold alloys often contain silver and copper. Since gold jewelry is normally made of alloyed gold rather than pure gold, it can also tarnish. The lower the karat value of the gold jewelry, the more likely it is to tarnish. Since jewelry gold is not pure, a variety of chemical products may discolor or dissolve it. A few of these products and their effect on gold and silver alloys are as follows:

♦ **Chlorine**—it can pit and dissolve the metal, causing prongs to snap and mountings to break apart. Solder connections are especially vulnerable. Afterwards, it might appear as if you've been sold defective or fake gold jewelry. Therefore, avoid wearing gold or silver jewelry in swimming pools or hot tubs that have chlorine disinfectants, and never soak it or clean it with bleach.

♦ **Lotions and cosmetics**—besides leaving a film on the jewelry piece, they may tarnish it, especially if it's made of 10K gold or silver. If possible, put your jewelry on last, after applying make-up and spraying your hair.

Many cosmetics are composed of minerals that are actually harder than gold. When these minerals are rubbed against your gold jewelry, very small particles of gold are removed from the jewelry and deposited on your skin, causing dark stains. This phenomenon, called black dermographism, is explained in the October 10, 2000 issue of the *Journal of Chemical Education* by Barbara Kebbekus.

♦ **Perm solutions**—they have a tendency to turn 10K gold and low-karat solder joints dark brown or black.

♦ **Some medications**—they may cause a chemical reaction in certain people. This can make their skin turn black when it comes into contact with the gold alloy.

♦ **Polishing compounds**—they can blacken your skin if they remain on the metal. Polishing cloths sold in jewelry stores may contain a mild abrasive for shining the metal. When using these cloths, be sure to wash or wipe the metal thoroughly afterwards.

Miscellaneous Tips

♦ If possible, avoid wearing jewelry while participating in contact sports or doing housework, gardening, repairs, etc. The mounting can be damaged, and stones can be chipped, scratched and cracked. During rough work, if you want to wear a ring for sentimental reasons or to avoid losing it, wear protective gloves. Hopefully, your ring has a smooth setting style with no high prongs.

♦ When you place jewelry near a sink, make sure the drains are plugged or that it's put in a protective container. Otherwise, don't take the jewelry off.

♦ Clean your jewelry on a regular basis. Then you won't have to use risky procedures to clean it later on.

♦ Don't remove rings by pulling on any of their gemstones. Instead, grasp the metal ring portion. This will help prevent the stones from coming loose and getting dirty.

♦ Occasionally check your jewelry for loose stones. Shake it or tap it lightly with your forefinger while holding it next to your ear. If you hear the stones rattle or click, have a jeweler tighten the prongs.

♦ Avoid exposing your jewelry to sudden changes of temperature. If you wear it in a hot tub and then go in cold water with it on, the stones could crack or shatter. Also, keep jewelry away from steam, hot pots and hot ovens in the kitchen.

♦ Take a photo of your jewelry (a macro lens is helpful). Just lay it all together on a table for the photo. If the jewelry is ever lost or stolen, you'll have documentation to help you remember and prove what you had. Expensive jewelry should be documented and appraised by a professional jewelry appraiser. You can find appraisal information in my *Gem & Jewelry Pocket Guide, Diamond Handbook, Ruby, Sapphire & Emerald Buying Guide, and Exotic Gems books.* My website **www.reneenewman.com** also has a list of independent jewelry appraisers and appraisal organization links. Simply click on the appraisers link.

♦ About every six months, have a jewelry professional check your rings for loose stones or wear and tear on the mounting. Many jewelers will do this free of charge, and they'll be happy to answer your questions regarding the care of your jewelry.

Bibliography

Books

Allen, Gina. *Gold!* New York: Thomas Y. Crowell, 1964.

Bovin, Murray. *Jewelry Making*. Forest Hills, NY: Bovin Publishing, 1967.

Branson, Oscar T. *What You Need to Know About Your Gold and Silver*. Tucson, AZ: Treasure Chest Publications, 1980.

Brod, I. Jack. *Consumer's Guide to Buying and Selling Gold, Silver, and Diamonds*. Garden City, NY: Doubleday, 1985.

Burkett, Russell. *Everything You Wanted to Know about Gold and Other Precious Metals*. Whittier, CA: Gem Guides Book Co., 1975.

Cavelti, Peter C. *New Profits in Gold, Silver & Strategic Metals*. New York: McGraw-Hill, 1985.

Cullen, V. Alexander. *Gold and Silver Scrap Dealers Handbook*, 2011.

Edwards, Rod. *The Technique of Jewelry*. New York: Charles Scribner's Sons, 1977.

Garrett, Jeff & Bressett, Kenneth. *Gold: Everything you need to know to Buy and Sell Today*. Atlanta: Whitman Publishing, LLC, 2010.

Gemological Institute of America. *Gold & Precious Metals Course*.

Gemological Institute of America. *Jewelry Repair Workbook*.

Gemological Institute of America. *Jewelry Sales Course*.

Gemological Institute of America. *Jewelry Essentials*

Goldsmiths Company & Johnson Matthey, *Palladium Technical Manual, UK Edition*.

Goldemberg, Rose Leiman. *Antique Jewelry: A Practical and Passionate Guide*. New York: Crown Publishing Co., 1976.

Green, Timothy. *The Gold Companion*. London: Rosendale Press, 1991.

Jarvis, Charles A. *Jewelry Manufacture and Repair*. New York: Bonanza, 1979.

Marcum, David. *Fine Gems and Jewelry*. Homewood, IL: Dow Jones-Irwin, 1986.

Maloney, Michael. *Guide to Investing in Gold and Silver*. New York: Business Plus, 2008.

McCreight, Tim. *The Complete Metalsmith: An Illustrated Handbook*. Worcester, MA: Davis Publications, 1991.

McCreight, Tim. *Jewelry: Fundamentals of Metalsmithing*. Madison, WI: Hand Books Press, 1997.

McCreight, Tim (editor). *Metals Technic.* Cape Elisabeth, Maine: Brynmorgan Press, 1992.

McGrath, Jinks. *The Encyclopedia of Jewelry Making Techniques.* Philadelphia: Running Press, 1995.

Merton, Henry A. *Your Gold & Silver.* New York: Macmillan, 1981.

Metal Man, *Secrets of a Successful Gold Buyer.* Be a Gold Buyer, Intl, 2010.

Miller, Anna M. *Gems and Jewelry Appraising.* New York: Van Nostrand Reinhold, 1988.

Miller, Anna M. *Illustrated Guide to Jewelry Appraising.* New York: Van Nostrand Reinhold, 1990.

Miller, Judith. *Costume Jewelry: The Complete Visual Reference and Price Guide:* London, 2003.

Miguel, Jorge. *Jewelry: How to Create Your Image.* Dallas: Taylor Publishing Co. 1986.

Morton, Philip. *Contemporary Jewelry.* New York: Holt, Rinehart, and Winston, 1976.

Newman, Renée, *Diamond Ring Buying Guide: 7th Edition.* Los Angeles: International Jewelry Publications, 2008.

Newman, Renée, *Jewelry Handbook:* Los Angeles: International Jewelry Publication, 2007.

Newman, Renée, *Gem & Jewelry Pocket Guide.* Los Angeles: International Jewelry Publications, 2006.

Newman, Renée, *Gold & Platinum Jewelry Buying Guide.* Los Angeles: International Jewelry Publications, 2000.

Ostier, Marianne. *Jewels and the Woman: The Romance, Magic, and Art of Feminine Adornment.* New York: Horizong Press, 1958.

Pitman, Ann Mitchell. *Inside the Jewelry Box. A Collector's Guide to Costume Jewelry.* Paducah, KY: Collector Books, 2004.

Penny Proddow, Marion Fasel. *With This Ring: The Ultimate Guide to Wedding Jewelry.* New York: Bulfinch, Press, 2004.

Platinum Guild International, *Platinum Starter Kit: Bench Companion,* 2012

Revere, Alan, *Professional Goldsmithing.* New York: Van Nostrand Reinhold, 1991.

Richards, Alison. *Handmade Jewelry.* New York: Funk & Wagnalls, 1976.

Ruhle-Diebener-Verlag. *Practical Platinumsmith: 3rd Edition.* Stuttgart: Ruhle-Diebener-Verlag, 1995.

Sarett, Morton R. *The Jewelry in Your Life.* Chicago: Nelson-Hall, 1979.

Sisk, Gerald D. *Guide to Gems & Jewelry.* Knoxville, TN: American's Collectibles Network, Inc., 2011.

Smith, Ernest. *Working in Precious Metals.* Colchester, England: N. A. G. Press Ltd, 1933.

Sprintzen, Alice. *Jewelry: Basic Techniques and Design.* Radnor, PA: Chilton, 1980.

Sutherland, C. H. V. *Gold: Its Beauty, Power and Allure*. New York: McGraw-Hill, 1969.

Untracht, Oppi. *Jewelry Concepts & Technology*. New York: Doubleday, 1982.

Von Neumann, Robert. *The Design and Creation of Jewelry*. Radnor, PA: Chilton, 1972.

Whetstone, William; Niklewicz, Danusia & Matula, Lindy. *World Hallmarks: Volume 1*. San Francisco: Hallmark Research Institute, 2010.

Magazines

Canadian Jeweler. Toronto, ON, Style Communications Inc.

Gem & Jewellery News. London: Gemmological Association of Great Britain (Gem A).

Jewelers Circular Keystone. New York, NY, Reed Business Information.

Jewellery Business. Richmond Hill, ON: Kenilworth Media, Inc.

Jewelry News Asia. Hong Kong: CMP Asia Ltd.

Jewellery Review. Hong Kong: Brilliant Art Group.

Lapidary Journal Jewelry Artist. Loveland, CO, Interweave Press LLC.

MJSA Journal. Providence, RI, Manufacturing Jewelers & Suppliers of America.

Metalsmith. Princeton, NJ: Society of North American Goldsmiths.

National Jeweler. New York, NY, VNU Business Publications.

Southern Jewelry News. Greensboro, NC: *Southern Jewelry News*.

Miscellaneous (Articles, Catalogues, Brochures, etc.)

A & A Jewelry Supply Catalogue.

Consumer and Corporate Affairs Canada. "A Guide to the Precious Metals Marking Act and Regulations."

David H. Fell & Company precious metals guide.

David H. Fell & Company, "Precious Metals Recycling 10."

Federman, David. "Electroforming: Big, bold and light." Modern Jeweler, p. 65-66, January, 1993.

GIA and the World Gold Council. "The Gold Seminar Handbook."

Grobet USA Catalogue.

Manchanda, Dippal. "Testing Gold and Other Nobel Metals by 'Touch Acid Technique.'" The Laboratory at the Birmingham Assay Office.

Johnson Matthey, "Platinum Jewelry Products."

Mercer, Meredith E. "Methods for Determining the Gold Content of Jewelry Metals." Gems & Gemology, p. 222-233, Winter 1992.

Metallix Newsletter

Platinum Guild International. "Talking Platinum."

"Platinum Alloys and Their Application in Jewelry Making" by the Platinum Guild International USA, written by Jurgen Maerz, Director of Technical Education with assistance from Taryn Biggs & Stefanie Taylor of Mintek, South Africa, Johnson Matthey, Engelhard-Clal, Imperial Smelting, C. Hafner Co. and Techform, 1999.

Reactive Metals Studio Inc. Online Catalogue, Clarkdale, Az.

Rio Grande *Gems & Findings* Catalogue.

Rubin & Son. *Supplies and Equipment for the Jewelry-Diamond Trade, Gemological Instruments.*

Stuller. *The Findings Book.*

Stuller. *The Mountings Book.*

Swest Inc. *Jewelers' Findings & Stones & Metals.*

Index

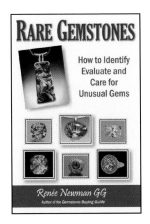

Rare Gemstones

Provides concise information on the identification properties, geographic sources, treatments, imitations, durability, uses, value factors and pricing of non-traditional gems. High quality photos show the different colors, cutting styles and varieties of each gem and give ideas on how each can be used creatively in jewelry.

"Rare Gemstones is **a fascinating insight into the latest and more unusual gemstones that** are now finding their way into designer jewellery . . . over 60 of the lesser-used gem materials have been selected, documented and presented in a highly visual way. Some of the stones covered have limited use due to hardness or durability factors, but these limitations are all addressed in the various sections. Where this book differs from others is in the extensive use of photographs of rough, cut and fashioned gemstones, as well as gem-set jewellery, showing that not only can these rare gemstones be used in various settings, but also that they are available today. Rough crystals and matrix are also illustrated in different settings including, for example, wrapped cobaltocalcite pendants and earrings, amongst others . . . The information supplied on each stone is comprehensive . . . Whether a newcomer to the world of gemmology or an experienced collector, this volume has something for everyone and is highly recommended."

Gems & Jewellery, published by the Gemmological Association of Great Britain

"Even if you're not in the market for rare gems, this book features hundreds of full-color photos of astonishing stone formations and the mouth-watering fine jewelry made with them. Can you say inspiration?" *Bead & Button Magazine*

"I wish all of the books that appraisers need to keep on our reference shelves were just like this one. *Rare Gemstones* is a treasure. **Like all of Renee Newman's books, it is impeccably organized, beautiful and complete.** This book is intended to be the companion to Renée's book *The Gemstone Buying Guide* . . . Probably the most beneficial aspect of *Rare Gemstones* for the appraiser is having all of this information put together in one affordable, easy to understand book, especially considering the newness of the information. I recommend that all appraisers buy this book."

The Jewelry Appraiser, Published by the National Association of Jewelry Appraisers

"This compact informative little book inspires one to look at different gemstones that one would not normally have considered as jewellery items and reassures you that with the right care (which is discussed in detail), they can be enjoyed just as much as fragile gems that are already widely accepted, such as pearls and amber. The book would be **invaluable to gemmology students studying 'Lesser Known Gemstones' and to anyone interested in rare gemstones."** *Australian Gemmologist*

137 pages, 482 photos, 6" by 9", ISBN 978-0929975-46-7, US$19.95

Other Books by RENÉE NEWMAN

Graduate Gemologist (GIA)

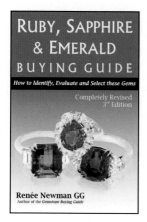

Ruby, Sapphire & Emerald Buying Guide

How to Identify, Evaluate & Select these Gems

An advanced, full-color guide to identifying and evaluating rubies, sapphires and emeralds including information on treatments, grading systems, geographic sources, fakes, synthetics, lab reports, appraisals, and gem care.

"Enjoyable reading . . . profusely illustrated with color photographs showing not only the beauty of finished jewelry but close-ups and magnification of details such as finish, flaws and fakes . . . Sophisticated enough for professionals to use . . . highly recommended . . . Newman's guides are the ones to take along when shopping."
Library Journal

"Solid, informative and comprehensive . . . dissects each aspect of ruby and sapphire value in detail . . . a wealth of grading information . . . a definite thumbs-up!"
C. R. Beesley, President, American Gemological Laboratories, *JCK Magazine*

187 pages, 280 photos, 267 in color, 6" by 9", ISBN 978-0929975-41-2, US$19.95

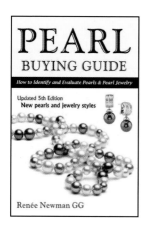

Pearl Buying Guide

How to Evaluate, Identify, Select and Care for Pearls & Pearl Jewelry

"Copious color photographs . . . explains how to appraise and distinguish among all varieties of pearls . . . takes potential buyers and collectors through the ins and outs of the pearl world."
Publisher's Weekly

"An indispensable guide to judging [pearl] characteristics, distinguishing genuine from imitation, and making wise choices . . . useful to all types of readers, from the professional jeweler to the average patron . . . **highly recommended."**
Library Journal

"A well written, beautifully illustrated book designed to help retail customers, jewelry designers, and store buyers make informed buying decisions about the various types of pearls and pearl jewelry."

Gems & Gemology

154 pages, 321 photos, 6" by 9", ISBN 978-0929975-44-3, US$19.95

For more information, see **www.reneenewman.com**

Exotic Gems, Volume 2

How to Identify and Buy Alexandrite, Andalusite, Chrysoberyl Cat's-eye, Kyanite, Common Opal, Fire Opal, Dinosaur Gembone, Tsavorite, Rhodolite & Other Garnets

"The subtitle of this book says is all: . . . Not familiar with some of these stones? Your jewelry designs might be missing out. Like its predecessor, *Exotic Gems, Volume 1*, this handy guide helps you find, evaluate, buy and care for these stones with plenty of photos of finished jewelry."

Bead & Button, reviewed by Stacy Werkheis

"The best thing about this series is the focus - Newman manages to include enough information and photographs from the likes of John Koivula interspersed with fun facts, metaphysical properties, history and ancient lore and so many photos that everyone from Gemologists to Jewelry Lovers will relate, understand and treasure these books . . . My favorite thing about these books is that Newman is not afraid to cover many of the opaque gems that have so long been overlooked as well as sourcing the latest, newest, and the sometimes controversial, gems in the field. Thanks for another winner Renee!"

"A Fly on the Wall Views & Reviews," reviewed by Robyn Hawk

"Renee Newman never disappoints us, does she? She has written another outstanding book about gemstones that we can learn from, teach from, and recommend to our clients . . . I do think that this book will serve appraisers well."

The Jewelry Appraiser, reviewed by Kim Piracci

". . . those familiar with Newman's previous books will recognise the in depth, yet understandable style, catering to professionals and lay people alike . . . Chapter nine on common opal is a real eye opener, Newman deftly capturing the beauty found in an assortment of colours including Peruvian pink, Andean blue, green, yellow and even lime through excellent and abundant photographs. Interesting issues about classification and the term 'fire' are discussed in the chapter on fire opal. Opal treatments are analysed in depth.

The remainder of the book is devoted to the complex group of garnets about which gemmologists and mineralogists disagree when defining species and varieties. Gemmologists will find the technical information in this section most valuable. In the fascinating pages of photographs, we are shown how a master cutter cuts a garnet. The 'horsetail' inclusions in demantoid garnets, unlike inclusions in most other garnets, are desirable. We are also directed to publications offering more detailed information.

Imparting a wealth of information on gemstone evaluation, as usual, with tips on detecting imitations, synthetics and gem treatments, Newman always entertains with interesting anecdotes of history, geographic sources and metaphysical lore of gems. Be ready to be informed and entertained. Didn't know what the 'alexandrite effect' was or what comprises the interesting crew digging up Arizona garnets? You will now."

The Australian Gemmologist, reviewed by Carol Resnick

154 pages, 408 color photos, 6" x 9", ISBN 978-0-929975-42-9, US$19.95

Other Books by RENÉE NEWMAN

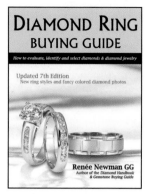

Diamond Ring Buying Guide

How to Evaluate, Identify and Select Diamonds & Diamond Jewelry

"**An entire course on judging diamonds in 156 pages of well-organized information**. The photos are excellent . . . Clear and concise, it serves as a check-list for the purchase and mounting of a diamond . . . another fine update in a series of books that are useful to both the jewelry industry and consumers."

Gems & Gemology

"**A wealth of information** . . . delves into the intricacies of shape, carat weight, color, clarity, setting style, and cut—happily avoiding all industry jargon and keeping explanations streamlined enough so even the first-time diamond buyer can confidently choose a gem."

Booklist

"Succinctly written in a step-by-step, outlined format with plenty of photographs to illustrate the salient points; it could help keep a lot of people out of trouble. Essentially, it is a **fact-filled text devoid of a lot of technical mumbo-jumbo.** This is a definite thumbs up!"

C. R. Beesley, President, American Gemological Laboratories

15 pages, 274 color & b/w photos, 7" X 9", ISBN 978-0-929975-40-5, US$18.95

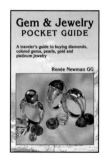

Gem & Jewelry Pocket Guide

Small enough to use while shopping locally or abroad

"**Brilliantly planned, painstakingly researched, and beautifully produced** . . . this handy little book comes closer to covering all of the important bases than any similar guides have managed to do. From good descriptions of the most popular gem materials (plus gold and platinum), to jewelry craftsmanship, treatments, gem sources, appraisals, documentation, and even information about U.S. customs for foreign travelers—it is all here. I heartily endorse this wonderful pocket guide."

John S. White, former Curator of Gems & Minerals at the Smithsonian *Lapidary Journal*

"**Short guides don't come better than this**. . . . As always with this author, the presentation is immaculate and each opening displays high-class pictures of gemstones and jewellery." *Journal of Gemmology*

154 pages, 108 color photos, 4½" by 7", ISBN 978-0929975-30-6, US$11.95

Available at major bookstores and jewelry supply stores

For more information, see **www.reneenewman.com**

Exotic Gems, Volume 1

This is the first in a series of books that explores the history, lore, evaluation, geographic sources, and identifying properties of lesser-known gems. *Exotic Gems, Volume 1* has detailed info and close-up color photos of mounted and loose tanzanite, labradorite, zultanite, rhodochrosite, sunstone, moonstone, ammolite, spectrolite, amazonite andesine, bytownite, orthoclase and oligoclase.

"Chapters including 'Price factors in a nutshell' will prove indispensable to novice buyers. The breadth of information on each stone, Renee's guide to choosing an appraiser, 288 vibrant photos and a bibliography also make this book a handy resource for seasoned collectors. We'll be watching for future installments of the *Exotic Gems* series." *Bead & Button*

". . . contains many many color photographs that cover the spectrum of subjects from mining locality shots to cutting to subtle color variations to the finished jewelry, as appropriate . . . A quick glance at the acknowledgments shows that a great deal of networking and editorial effort has gone into this book. If you want to buy one of the materials covered by this book, already have spent your money but want an appraisal, or are just plain interested in zultanite, I highly recommend *Exotic Gems Volume 1.*" *Rocks & Minerals*

154 pages, 288 color photos, 6" x 9", ISBN 978-0-929975-42-9, US$19.95

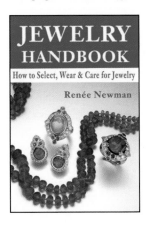

Jewelry Handbook
How to Select, Wear & Care for Jewelry

The *Jewelry Handbook* is like a Jewelry 101 course on the fundamentals of jewelry metals, settings, finishes, necklaces, chains, clasps, bracelets, rings, earrings, brooches, pins, clips, manufacturing methods and jewelry selection and care. It outlines the benefits and drawbacks of the various setting styles, mountings, chains, and metals such as gold, silver, platinum, palladium, titanium, stainless steel and tungsten. It also provides info and color photos on gemstones and fineness marks and helps you select versatile, durable jewelry that flatters your features.

"**A great introduction to jewellery** and should be required reading for all in the industry." Dr. Jack Ogden, CEO Gem-A (British Gemmological Association)

"**A user-friendly, beautifully illustrated guide,** allowing for quick reference to specific topics." *The Jewelry Appraiser*

"**Valuable advice for consumers and the trade**, specifically those in retail sales and perhaps even more for jewelry appraisers . . . An easy read and easy to find valuable lists and details." Richard Drucker GG, *Gem Market News*

177 pages, 297 color & 47 b/w photos, 6" x 9", ISBN 978-0-929975-38-2, US$19.95

Other Books by RENÉE NEWMAN

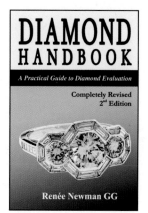

Diamond Handbook
A Practical Guide to Diamond Evaluation

Updates professionals on new developments in the diamond industry and provides advanced information on diamond grading, treatments, synthetic diamonds, fluorescence, and fancy colored diamonds. It also covers topics not in the *Diamond Ring Buying Guide* such as diamond grading reports, light performance, branded diamonds, diamond recutting, and antique diamond cuts and jewelry.

"**Impressively comprehensive**. . . . a **practical, well-organized and concisely written** volume, packed with valuable information. The *Diamond Handbook* is destined to become an indispensable reference for the consumer and trade professional alike."
Canadian Gemmologist

"The text covers everything the buyer needs to know, with useful comments on lighting and first-class images. No other text in current circulation discusses recutting and its possible effects . . . **This is a must for anyone buying, testing or valuing a polished diamond and for students in many fields**." *Journal of Gemmology*

186 pages, 320 photos (most in color), 6" x 9", ISBN 978-0-929975-39-9, US$19.95

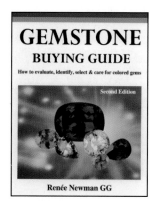

Gemstone Buying Guide
How to Evaluate, Identify and Select
Colored Gems

"Praiseworthy, **a beautiful gem-pictorial reference** and a help to everyone in viewing colored stones as a gemologist or gem dealer would. . . . One of the finest collections of gem photographs I've ever seen . . . If you see the book, you will probably purchase it on the spot."
Anglic Gemcutter

"**A quality Buying Guide** that is recommended for purchase to consumers, gemmologists and students of gemmology—irrespective of their standard of knowledge of gemmology. The information is comprehensive, factual, and well presented. Particularly noteworthy in this book are the quality colour photographs that have been carefully chosen to illustrate the text." *Australian Gemmologist*

"**Beautifully produced**. . . . With colour on almost every opening few could resist this book whether or not they were in the gem and jewellery trade."
Journal of Gemmology

154 pages, 281 color photos, 7" X 9", ISBN 978-0929975-34-4, US$19.95

Osteoporosis Prevention

" . . . a complete, practical, and easy-to-read reference for osteoporosis prevention . . . As the founding president of the Taiwan Osteoporosis Association, I am delighted to recommend this book to you."

Dr. Ko-En Huang, Founding President of TOA

"The author, Renée Newman has abundant experience in translating technical terms into everyday English. She writes this book about osteoporosis prevention from a patient's perspective. These two elements contribute to **an easy-to-read and understandable book for the public. To the medical professions, this is also a very valuable reference**."

Dr. Chyi-Her Lin, Dean of Medical College, Natl Cheng Kung Univ / Taiwan

"I was impressed with the comprehensive nature of *Osteoporosis Prevention* and its use of scientific sources. . . .The fact that the author has struggled with bone loss and can talk from personal experience makes the book more interesting and easy to read. Another good feature is that the book has informative illustrations and tables, which help clarify important points. I congratulate the author for writing **a sound and thorough guide to osteoporosis prevention**."   Ronald Lawrence MD, PhD
Co-chair of the first Symposium on Osteoporosis of the National Institute on Aging

" . . . **clarifies the inaccurate concepts from the Internet**. It contains abundant information and really deserves my recommendation."

Dr. Yung-Kuei Soong, The 6th President of Taiwanese Osteoporosis Association

"The book is written from a patient's experience and her secrets to bone care. This book is **so interesting that I finished reading it the following day** . . . The author translates all the technical terms into everyday English which makes this book so easy to read and understand."

Dr. Sheng-Mou Hou, Ex-minister, Dept. of Health / Taiwan

"**A competent and thoroughly 'reader friendly' approach to preventing osteoporosis**. Inclusive of information on how to: help prevent osteoporosis and broken bones; get enough calcium and other bone nutrients from food; make exercise safe and fun; retain a youthful posture; select a bone density center; get maximum benefit from your bone density exam; understand bone density reports; help seniors maintain their muscles and their bones; and how to be a savvy patient. *Osteoporosis Prevention* should be a part of every community health center and public library Health & Medicine reference collection . . ."

Midwest Book Review

"With great interest, I have read Renée Newman's *Osteoporosis Prevention* which provides complete and practical information about osteoporosis from a patient's perspective. . . . **a must-read reference for osteoporosis prevention**."

Dr. Tzay-Shing Yang, 3rd President of TOA, President of Taiwan Menopause Care Society

You can get free information about osteoporosis prevention, bone density testing and reports at: **www.avoidboneloss.com**

176 pages, 6" X 9", ISBN 978-0929975-37-5, US$15.95

Order Form

TITLE	Price	Quantity	Total
Gold, Platinum, Silver, Palladium & Other . . .	$19.95		
Rare Gemstones	$19.95		
Exotic Gems, Volume 2	$19.95		
Exotic Gems, Volume 1	$19.95		
Ruby, Sapphire & Emerald Buying Guide	$19.95		
Gemstone Buying Guide	$19.95		
Diamond Handbook	$19.95		
Pearl Buying Guide	$19.95		
Jewelry Handbook	$19.95		
Diamond Ring Buying Guide	$18.95		
Gem & Jewelry Pocket Guide	$11.95		
Osteoporosis Prevention	$15.95		
		Book Total	
SALES TAX for California residents only		**(book total x $.09)**	
SHIPPING: USA: first book $4.00, each additional copy $2.00 Canada & Mexico - airmail: first book $12.00, ea. addl. $5.00 All other foreign countries - airmail: first book $17.00, ea. addl. $5.00			
TOTAL AMOUNT with tax (if applicable) and shipping (Pay foreign orders with an international money order or a check drawn on a U.S. bank.)		**TOTAL**	

Available at major book stores or by mail. For quantity orders e-mail: intljpubl@aol.com

Mail check or money order in U.S. funds

To: International Jewelry Publications
P.O. Box 13384
Los Angeles, CA 90013-0384 USA

Ship to:

Name_____

Address_____

City_____ State or Province_____

Postal or Zip Code_____ Country